Wireless Internet Telecommunications

K. Daniel Wong

ARTECH
HOUSE

BOSTON | LONDON

www.artechhouse.com

Library of Congress Cataloging-in-Publication Data
Wong, K. Daniel.
  Wireless Internet telecommunications / K. Daniel Wong.
    p. cm.—(Artech House mobile communications series)
  Includes bibliographical references and index.
  ISBN 1-58053-711-1 (alk. paper)
    1. Wireless Internet.   2. Wireless communication systems.   I. Title.   II. Series

  TK5103.4885.W5718 2005
  004.67'8—dc22

                                                                    2004058540

British Library Cataloguing in Publication Data
Wong, K. Daniel
  Wireless Internet telecommunications.
  —(Artech House mobile communications series)
  1. Wireless Internet   2. Computer network protocols   3. Wireless communication
  systems
  I. Title
  621.3'8212

ISBN 1-58053-711-1

Cover design by Yekaterina Ratner

International Standard Book Number: 1-58053-711-1

10 9 8 7 6 5 4 3 2 1

Contents

CHAPTER 6

Mobility Management   91

CHAPTER 7

QoS   119

Preface

Many people try to predict the future. Weather forecasters do it. Financial analysts do it. Science fiction writers do it. We make and change travel plans based upon weather forecasts. We buy and sell stocks based upon predictions of future performance from financial analysts. Science fiction writers have successfully predicted advances like travel to the moon, robots, and communication satellites. Success rates strongly depend on the bases upon which the predictions are made. When weather forecasters obtain good information from satellites and other sources, or when financial analysts do their homework with accurate information on fundamentals, they could be quite successful in predicting the future. When science fiction writers have a good grasp of current scientific knowledge and let their imaginations roam within limits, they could be remarkably successful as well.

In writing this book, I am trying to predict the future. I am predicting that wireless communications and the Internet are both going to continue growing healthily. I am predicting that the intersection of wireless and the Internet is going to increase in significance. The bases for making these predictions are very stable and strong. The demands for wireless and for the Internet continue to grow. Meanwhile, the supporting technologies are experiencing exponential growth, as a variety of new wireless technologies are hitting the market faster than in the past, and Internet standards are growing in number faster than ever before. The range of applications and services that need to be supported by wireless and the Internet has broadened, and the technology has been challenged to keep up with the requirements. There is tremendous interest and market potential in the wireless Internet. What makes it even more exciting is that the supporting technologies, to support mobility, differentiated service quality, and security, to transport multimedia traffic over IP, and so on, are only now becoming mature and coming together to work together in practical systems.

This book should be useful for professionals in the telecommunications field (e.g., network architects, system engineers, and network software engineers, including both individual contributors and management). However, it should also be helpful for professors, consultants, and short-course instructors, as well as supplementary reading material for senior-level and master-level students in electrical engineering and computer science interested in finding out what wireless IP technology is about and how it works.

Acknowledgments

Artech House's anonymous reviewer provided many good suggestions and pointed out holes and other weaknesses in the first draft of the manuscript. Mr. Hon San Wong read through the entire manuscript and provided numerous editorial corrections and suggestions for improvement of the explanations, as well as assisted with a few of the figures. He also helped compile the index and proofread some of the galley proofs. Ms. Bhawani Selvaretnam helped to proofread the page proofs. Mr. Kok Seng Wong reviewed many of the manuscript chapters. Dr. Gary Chan, Dr. Mooi Choo Chuah, and Ms. Yin Kia Chiam each reviewed a couple of chapters. The feedback from the reviewers was tremendously helpful for me, and for that I thank them. I thank the professional editorial staff at Artech House, especially Mrs. Christine Daniele, Ms. Barbara Lovenvirth, and Ms. Rebecca Allendorf for their support and advice during the book proposal, writing, and production process.

I am grateful to Dr. Melbourne Barton, Dr. Russell Hsing, Dr. Ken Young, and Dr. Li Fung Chang for giving me the opportunity to work on projects in various areas in wireless and IP while in Telcordia Technologies. Thanks are also due to other colleagues in Telcordia Technologies, including Dr. Vijay Varma, Mr. Ashutosh Dutta, Dr. John Lee, and many others, for numerous discussions on relevant topics that deepened my understanding of the issues. I thank Dr. Nor Adnan Yahaya, Dr. Tajul Arus, and the rest of the staff at the Malaysia University of Science and Technology, for creating a working environment that was conducive for the writing of this book.

I am thankful to Jesus Christ for a wonderful family, and to my family for their support, encouragement, and understanding while I was working on this book.

CHAPTER 1

Introduction

I have written this introductory chapter to communicate why I am passionate about wireless Internet telecommunications. I hope this chapter will motivate you to read the rest of the book, by introducing an exciting vision of the future of communications, and starting on the road to explaining technologies to realize this vision. In this chapter, I also preview the rest of the book and explain the scope of coverage. Some of the terms introduced in this chapter may not be familiar to you, but they will be explained in due course.

1.1 An Exciting Future

Between the late 1970s and early 1990s, developments and inventions in two different areas—wireless and the Internet—were going on roughly at the same time, each in its own world. Especially in the 1990s, each area grew and blossomed tremendously. Second generation (2G) digital wireless personal communications phones were acquired by significant portions of the general population, becoming almost a standard personal accessory by the end of the decade. Meanwhile, the Internet also became a part of popular culture, due in large part to the success of the World Wide Web. Today, young people in their teens and twenties are among the most comfortable users of wireless and the Internet, and the next generation may be at the forefront of future advances in these areas. Consider the observation regarding contemporary college students that "they surf the Web during class, get to know their soul mates through instant messaging and talk on cell phones while biking across the Quad. Make way for the technology natives," and the statement, "I don't understand how people used to live before the Internet existed" [1].

Today, the Internet and wireless are converging towards an exciting future. This can be partly explained by the recent convergence of communications and computing technologies. However, it is also indicative of the possibilities that could be obtained by merging the features of the Internet and of wireless communications. Many of these features, related to quality of service (QoS), security, mobility, and multimedia traffic support, are only recently maturing, so we are on the verge of a revolutionary leap.

For example, we consider an imaginary scenario set sometime in the future involving a family of four, with Alice and Bob the parents and Charles and Diana

1

the children. Each member of the family has a personal communications device (we call it a "communicator" for purposes of this example) that he or she carries around everywhere. In the morning, when she wakes up, Alice finds a video message or e-mail on her communicator. It has arrived overnight through the fourth generation (4G) wireless network and it's from her secretary, reminding her of an important meeting today. She rushes off to work, while Bob prepares breakfast with a few key presses on his communicator—he has previously asked it to remember the default services for breakfast preparation, which include requests to the toaster to toast five slices of bread, the refrigerator to prepare four glasses of milk, the coffeemaker to brew a pot of coffee, and the kitchen TV set to tune in to the morning news. These communications happen over the home wireless network based on wireless local area network (LAN) or similar technology. Meanwhile, the refrigerator finds that it is running low on milk and eggs, and places an order with a local grocery store.

While Bob, Charles, and Diana are having breakfast, a TV commercial advertises discount tickets to a concert. Charles is interested to know more and speaks to his communicator about his interest. It connects him to a Web page. After browsing a few minutes, he indicates that he wishes to speak to a salesperson about the concert, and he is connected. Charles has not specified a preference for a voice-only or a video call, but in this case, the other side has a preference for voice-only, so the call is voice-only. A secure channel for this Voice over Internet Protocol (VoIP) traffic is set up, and the salesperson's identity is authenticated, allowing Charles to feel comfortable providing his credit card number. Diana meanwhile uses her communicator to check her buddy list of friends. Each of them has their own specifications for their current availability (e.g., available for voice calls and to receive images, but not open for video calls), which are color-coded for convenience. The friends have different levels of availability because their communicators and/or subscriptions have different sets of capabilities, and also because they can choose to limit their availability at any time. Diana has also noticed that the video quality while communicating with her richer friends is usually better than the video quality while communicating with her poorer friends. This is because her richer friends are subscribing to costlier packages that provide higher qualities of service.

Meanwhile, Alice is driving to work and mentally rehearsing the points she wishes to bring up at the meeting. She finds the most convenient fast-food store from which to pick up some breakfast, using a store locator service through her communicator. The store locator service provides the answer based on Alice's current location. After breakfast, her friend Karen calls, wanting to initiate a video call with Alice. Alice's car is equipped to display the video images on the windshield (a "heads-up display"), so she can drive while having the video call, but she does not want to be disturbed at the moment. She requests her communicator to schedule a time after 3 p.m. when she can talk with Karen. She does not specify a particular time, except that it must be a time slot when she does not have something else already scheduled. Her communicator checks her calendar, then communicates with Karen's communicator to make the arrangements. Meanwhile, Alice needs to check some points in an important document. She requests her communicator to download the document and display pages nine and ten on her heads-up display. Because Alice has a high-priority business subscription, the large file transfers almost instantaneously. Also, as a fringe benefit, the light classical music playing in the background on her personalized radio service provides crystal-clear sound. All this is happening as the car is moving at highway speeds along a highway and smoothly handing off between base stations of the 4G mobile wireless system, so as far as Alice is concerned, the connection is seamless even though she is moving.

Of course, this is only a projection of how the future might look. There could always be new ideas (technologies, applications, and so forth), even some disruptive technologies, that result in changes of direction, with different emphases. However, this scenario and others like it are good guides for shaping our thinking, and they provide targets for which to aim.

1.2   Requirements

This vision of the future requires a number of technological breakthroughs, some of which have already been made and are available today. Underlying these requirements is the cost constraint—the features need to be available at reasonable costs, to avoid being merely technology without commercial value. Furthermore, the relationship between the new services and applications, and the supporting technological innovations, is not so much a case of our having a clear idea exactly what the services and applications are and what they need, and then just solving the technical problem. Rather, the relationship is more of a continual iterative process, as shown in Figure 1.1. For example, in the early days of the Internet, when the technology was being designed, wireless services and applications were beyond the horizon. Therefore, later on, with the coming together of wireless and Internet protocol (IP), IP needed to be enhanced to support mobility. This in turn has been fueling further developments of services and applications like location-based services that can take

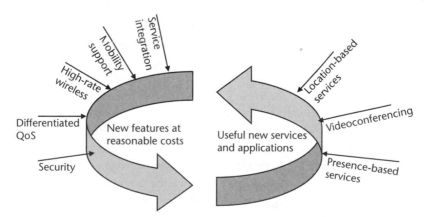

Figure 1.1   Getting there, through continual interactions involving new underlying features and the new services and applications they support.

advantage of mobility and other recent technological advancements. Similarly, even as different services and applications have driven the need for better and varied levels and types of security, the technology has matured to the point that consumer financial transactions are increasingly being conducted over the Internet, and other new and innovative services and applications are being tried out as well.

1.2.1 Technological Requirements

The requirements to realize this future include:

1. High level of service integration;
2. Advanced service enabling software technologies;
3. High-rate, reliable wireless communications;
4. Mobility support;
5. Supporting network infrastructure providing:
   - Service differentiation;
   - Secure communications.

Except for point 1, most of these points are explored at length in subsequent chapters in this book, so we discuss point 1 here and only briefly introduce the rest of these points in this section.

The vision of the future wireless Internet presented in Section 1.1 requires a high level of service integration. When services like voice and the Internet are highly integrated, a user can click on a link on a commercial Web page to initiate a call to customer service, or a user can pick up voice and e-mail messages from either a phone or mail program, or a user can dial the same number and have it simultaneously ring both a regular phone and the "Internet phone" on the called party's laptop. While such integrated services may have particular names like computer-telephony integration (CTI), unified messaging, or call forking, depending on what services are being integrated, the underlying idea is the integration of services that used to be separate.

Going one step further, we can argue for greater integration of communications systems, not just integration of services. This is sometimes called convergence. In fact, different convergences are occurring, including convergence of computing and communications, convergence of wireless and the Internet, and convergence of communications systems. Here I am referring to the convergence of communications systems. Traditionally, engineers have designed different communication systems separately, each system with its own intended services that are provided in its own way to its intended users. For example, there is the telephone system that provides mainly telephony services, and there are various data network systems that provide mainly data networking services. Such systems are also called "stovepipe" solutions, and they have their own way of solving the various subproblems like networking and transport. Convergence of communications systems is about knocking down the walls between the stovepipe solutions.

Clearly, the converged systems would need to be multiservice systems, which are systems that provide multiple services, such as voice and video telephony and data communications services, instead of just one or a few services, as with the stovepipe systems. A major advantage of convergence is the greater potential for service integration. A tradeoff is that each stovepipe system may be highly optimized for the few services that it provides, whereas the converged network may not be able to be so optimized for particular services, since it needs to support multiple services. For example, as traditionally circuit-switched systems for voice telephony converge with traditionally packet-switched systems for data like the Internet, the converged network could be either circuit-switched or packet-switched. Although circuit switching is generally more optimal for voice telephony, packet switching may be chosen for the converged network. Nevertheless, research is in progress in voice over packet switching (discussed in Chapter 4), to make the best of a less-optimal solution.

Moving on to the next requirement, I believe that the software and software architectures that enable services in the multiservice converged networks will need to be more sophisticated than in stovepipe systems, and flexible enough to support innovative integrated services and applications. Traditionally powerful concepts in the architecting of complex systems can be applied—concepts like abstraction and layering. Thus we are seeing middleware concepts like open systems access (OSA) emerge (discussed in Chapter 10), allowing application software to make use of features like location information without needing to understand how the information is obtained. Instead, lower-layer features are abstracted for use by higher layers.

The wireless medium is notoriously difficult to work with. Signal strength can fluctuate greatly due to fundamental problems like the multipath phenomenon, shown in Figure 1.2 (the figure illustrates multipath between a transmitter and a cell phone, and between the same transmitter and a laptop). Multipath describes when the signal from a transmitter reflects off different objects and takes multiple paths to the receiver, potentially causing destructive interference to itself. It is therefore a

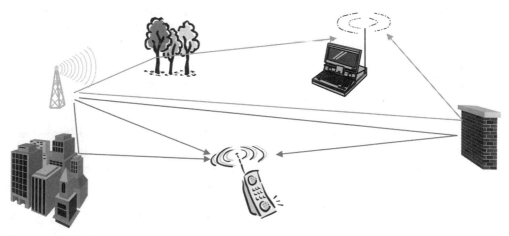

Figure 1.2   The fundamental problem of multipath in wireless.

great challenge to provide high data rates while at the same time ensuring highly reliable communications. How high are the data rates needed? How reliable should wireless communications be? Furthermore, the duration is also an issue—sustained high data rate wireless communications consumes tremendous amounts of resources. Wireless third generation (3G) systems provide higher data rates than their 2G counterparts, but these rates are not as high as the rates provided by wireless LAN systems. A number of technologies in the pipeline may provide even higher data rates for the fourth generation (4G) systems (discussed in Chapter 13).

Wireless provides for freedom of movement of communications devices. However, that adds challenges for the network to keep track of the location of these devices and to provide for communications even as the devices move across points of attachment to the network. Mobility management (see Chapter 6) is needed. A variety of schemes for QoS differentiation (see Chapter 7) have been proposed, and work is continuing to ensure that wireless IP telecommunications systems will have the necessary service quality foundations. Security is also a serious concern, especially given the additional challenges that are faced in wireless environments (compared with wireline environments). The security aspects (see Chapter 8) of these systems are closely tied with mobility aspects, because the need to handle mobility well is one of the reasons why maintaining security is challenging in wireless Internet telecommunications.

1.3   Preview

Wireless communications and Internet-based communications have been growing rapidly in recent years. In the early years of wireless cellular systems, most of the interest in wireless was focused on circuit-switched voice communications. However, the Internet has been growing, and the volume of packet-switched data traffic along with it. Moreover, 802.11-based wireless LANs have been growing in popularity, and often are used as an extension to the wired Internet. Therefore, it is making increasing economic sense for voice and data to share a common packet-switched infrastructure, with IP-based packet switching as the natural candidate for most cases. It is important to note that both 3GPP and 3GPP2 (standards groups to design systems for 3G cellular wireless and beyond, which will be discussed later in the book) are moving towards the all-IP wireless network concept.

Since many readers may be familiar with either IP or wireless, but not necessarily both, Chapters 2 and 3 provide brief tutorials on IP and Internet concepts, and on wireless networks, respectively. The coverage in these chapters is aimed at bringing the newcomer (to either IP or wireless) up to speed quickly, while at the same time touching on issues that relate to the expositions in later chapters. If you are a newcomer to either IP or wireless, you should be able to read and understand the subsequent chapters after reading the introductions in Chapters 2 and 3. Nevertheless, the references provided in these chapters will be helpful for further study and background reading in IP and wireless technology, respectively.

The wireless Internet is expected to support a variety of applications, some new and some evolved from existing wireless or Internet applications. The communication of multimedia content will be featured in many of these applications. Thus, after introducing wireless and the Internet, the book will discuss technologies related to communicating multimedia over IP. There are a variety of recent advances in technologies for packet-switched voice, video, and other multimedia. Protocols like Real-time Transport Protocol (RTP) have been developed to handle multimedia transport, addressing issues like synchronization of different components of real-time multimedia streams. Session Initiation Protocol (SIP) is gaining popularity as a flexible protocol for session control. Multimedia transport issues will be discussed in Chapter 4, whereas session control with SIP will be discussed in Chapter 5.

The three main challenges in wireless networking, viewed from a broader perspective (i.e., not just in the context of multimedia traffic), are mobility, QoS, and security. Mobility management is explored in Chapter 6. Chapter 7 discusses QoS. Chapter 8 discusses security. Meanwhile, the Internet Protocol itself is being upgraded to incorporate various features that will be more conducive for its use in future networks, including wireless IP networks. As such, topics including mobility support and support for a very large address space are included in Internet protocol, version 6 (IPv6), which will be introduced in Chapter 9. The services and applications driving any network technology are the keys to success. Recent advances in thinking about and understanding such topics as services and applications and middleware will be discussed in Chapter 10.

Wireless mobile systems are evolving from 2G systems to 3G systems, which will happen in phases. As we move past the first couple of phases, 3G systems will move towards being IP-based. The IP multimedia subsystem (IMS) in the Universal Mobile Transmission System (UMTS) developed by 3GPP will be introduced in Chapter 12, after we trace the evolution of the Global System for Mobile communications (GSM) to UMTS in Chapter 11. This will serve two purposes. First, of the wireless all-IP systems being developed, IMS in UMTS is the furthest along in development. Being on the cutting edge, IMS uses the latest Internet technologies, and IMS developments are also being fed back into the Internet Engineering Task Force (IETF), shaping the development of IPv6 and protocols like SIP. Therefore, understanding this system will be of great importance and interest to the reader of this book. Second, UMTS and IMS will be used to illustrate how the various components come together in an example of an all-IP system. This should help the reader to understand better how various aspects, such as session control signaling, QoS control, and security, can be put together in a real system. Where other alternatives (i.e., different from the actual choices 3GPP made in designing IMS) are possible, these will also be discussed.

While wireless IP telecommunications is still evolving, it is imperative to look forward at possibilities for future developments. This will help to impart to the reader a feeling for the evolving nature of the field, and some of the areas and topics the reader may encounter (and perhaps be actively working in) in the future. Therefore, Chapter 13 focuses on future developments and the elusive concept of 4G.

The overall book organization is shown pictorially in Figure 1.3, where the relationships of the topics to one another is shown, and where the corresponding book chapters are shown in parentheses.

1.4 Themes and Principles

In this section, I lay out some of the broad themes that you will see throughout the book. I also put forward some general principles that can help frame your thinking of the issues involved.

The themes are:

1. Convergence;
2. The bell-heads versus the net-heads;
3. Piecewise versus monolithic specifications;
4. Intelligence in the network versus intelligence in the edge.

The general principles are:

1. The design concept of a system or protocol can be quite different from its practical usage design for flexibility.
2. Pay due respect to the problem of transitioning from one technology to another.

In Section 1.2.1, we have already seen how pervasive the phenomenon of convergence in modern telecommunications is. At the broadest level, there is a

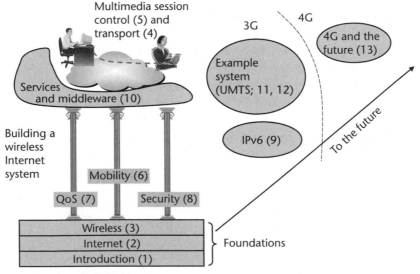

Figure 1.3 Organization of the book.

convergence of computing and communications, as communications increasingly support distributed computing, while software technologies are revolutionizing telecommunications devices. The convergence of wireless and the Internet is the subject of this book, and the convergence of communications systems looks set to continue into the future, greatly impacting the architecture of 4G wireless systems. Voice and video over IP (and more generally, multimedia over IP) is one of the most important applications of the wireless Internet. This application represents another example of convergence—voice and data used to be handled by different networks, but they are converging in the wireless Internet.

One of the oldest and most fascinating friendly (and sometimes not so friendly!) feuds in telecommunications is between the so-called bell-heads and net-heads. The bell-heads are the individuals who come from the traditional telephony world (or at least, who have such a frame of mind), where voice is king and networks are engineered for extreme reliability based on circuit switching. These persons are perhaps called bell-heads after the inventor of the telephone or the Bell companies in the United States. The net-heads are the individuals who come from the traditional data communications world, where packet switching instead of circuit switching has gained the upper hand. However, the feud is not just a debate between circuit switching and packet switching, or between more reliable and less reliable networks, but something of a holy war between different system design philosophies. Two major aspects of the differences in philosophies can be seen in the different positions net-heads and bell-heads tend to take on two themes.

On the theme of piecewise versus monolithic specifications, the net-heads tend to prefer piecewise specifications, whereas the bell-heads prefer monolithic specifications. By monolithic, we mean a well-integrated and complete specification. It may be best to illustrate the differences by example. The Internet, and its related protocols, is a prime example of a system that follows the piecewise specification model. Each protocol is designed for a specific purpose, and meant to do only that purpose, but do it well. On the other hand, the phone system is more of a monolithic system, for instance, in terms of being a stovepipe solution (as discussed earlier). While a monolithically specified system may be more robust and reliable, it may be less flexible to be used in different scenarios. On the other hand, the flexibility of the piecewise approach is not always a good thing, as it allows more scope for bad systems integration resulting in poorly performing systems.

Another philosophical area of incompatibility between bell-heads and net-heads has to do with where the intelligence in the network lies. The Internet purist sees a network that is stateless and "stupid" in the sense that it concentrates on forwarding packets rather than providing many services. Many services are best handled by the end points, a principle sometimes articulated as the "end-to-end principle" [2]. While this ideal might seem good, many useful and practical things in the Internet violate the ideal, putting intelligence and state in the network (e.g., NATs in Chapter 2, IPsec gateways in Chapter 8). On the other hand, with the phone network, the end points are typically stupid, and the network contains the intelligence and state for service provision. In an interesting and provocative article, Isenberg proclaimed the dawn of the stupid network, and that the days of the old-style "intelligent network" of the phone companies were coming to an end [3]. The article was published in 1998 and today, intelligent networks are still around—showing that things were perhaps not as simple as Isenberg made them out to be.

Of the two design principles listed above, we cannot understate the importance of flexible design. Given the rapid pace of change in technologies, the telecommunications industry protocols need to be adaptable to keep up with the latest developments and usage scenarios. For example, SIP (see Chapter 5) is a good example of a well-designed protocol that has grown in ways that its creators had not thought of initially, as it is now being applied in many scenarios with different requirements, such as in the IMS in UMTS (see Chapter 12). However, due to its flexibility and resilience, SIP is adapting well. The Internet itself is constantly being grown and pushed to handle requirements for which it was not initially designed (such as mobility, QoS, and security, as mentioned earlier). Similarly, wireless systems like GSM were originally designed primarily for mobile phone services, and they have needed to grow and adapt to handle more data services [thus, General Packet Radio Service (GPRS) was added], to provide location information on mobile devices, and to perform other functions. UMTS, the successor of GSM, is moving towards an all-IP wireless network concept, and its IMS will utilize the latest version of IP, namely IPv6. Also, wireless LAN has been found to be so useful and versatile that it is being proposed for usage in new scenarios such as for intervehicular communications (albeit with suitable modifications).

The other important principle is to respect the technology transition problem. Whether it is moving from 2G to 3G systems, or from Internet protocol, version 4, (IPv4) to IPv6, or from circuit-switched voice to packet-switched voice, there is the issue of how to migrate from the old solution to the new one. Planners have learnt to respect the power of incumbency (of the old systems), so new technologies often are designed for backward compatibility, perhaps able to interwork with the old technologies through gateways, packet encapsulations, or other means. Reasonable technology transition plans are also crucial. Sometimes, even with backward compatibility and interworking built in, the transition needs an external catalyst to get going. For example, one reason why the first UMTS system was deployed in Japan was that the 2G system was running out of capacity. Another example is that it may take the enforced utilization of IPv6 in large systems to get the momentum going. This may happen with the widespread adoption of the IMS that mandates use of IPv6.

We have discussed these themes and principles because they help us appreciate better the fast-paced developments in wireless Internet telecommunications, and to have a broad perspective on what is going on. After all, technology is often not created in a vacuum, but is part of the broader context of technology development and issues in that field. This applies to wireless Internet telecommunications too.

1.5 Scope

I have made difficult choices in deciding what to include in and exclude from this book. This book emphasizes a "how it works" approach, to give the reader a big-picture overview of how wireless Internet telecommunications works, by surveying the wireless Internet landscape. I hope that enough details are given that the reader who needs to pursue specific issues in more depth will be easily able to find and understand the appropriate material. Although quantitative performance analysis is not within the scope of this book, I hope that this book gives interested readers a strong enough grasp of the subject to research it further.

The subject of wireless is very broad, and this book focuses on the category of wireless personal communications and, more specifically, terrestrial wireless personal communications systems. We will be interested mainly in wireless systems that support mobility and handoffs between points of attachment to the network. Thus, we do not consider here wireless satellite communications, wireless broadcasting (such as TV or radio), and fixed wireless systems (such as point-to-point microwave systems and wireless local loop). Cordless phones are also out of our scope because they do not support handoffs. The wireless systems that support mobility and handoffs are infrastructure-backed wireless networks and infrastructureless wireless networks. Infrastructure-backed wireless networks have an infrastructure, typically wired, that supports the wireless links. Cellular networks are examples of this type. Infrastructureless wireless networks are self-contained wireless networks that do not need a wired infrastructure; they are also known as *multihop ad hoc networks*, or ad hoc networks for short. There are many differences in the challenges and issues for infrastructure-backed wireless networks and ad hoc networks. Thus, for a sharper focus, in this book I concentrate on infrastructure-backed wireless networks.

While a complete system includes the physical layer as well as application software and middleware, this book focuses on the networking aspects of wireless Internet telecommunications. Borrowing Deering's "protocol hourglass" analogy [4], we see that the IP works over many link- and physical-layer protocols and technologies, and at the same time there are many protocols at each of the higher layers. Figure 1.4 shows the IP hourglass and Figure 1.5 shows the areas of the hourglass the book addresses. Rather than discussing many applications in the wireless Internet, we focus on an important example, conversational[1] multimedia over IP, and show how it is supported over IP in the layered architecture.

This book does not go into detail regarding network management (such as configuration and provisioning), issues related to enterprise requirements (for robustness, billing, and commercial strength), and business aspects of the wireless Internet. Bringing in such issues would complicate the text unnecessarily. However, such

---

1. We say "conversational" to distinguish it from streaming multimedia. In the conversational case, two or more parties are involved in real time, whereas streaming is mostly one-directional. In Chapter 7, we point out that streaming is more tolerant of jitter than conversational traffic.

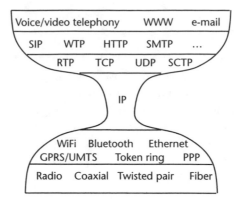

Figure 1.4 The IP hourglass.

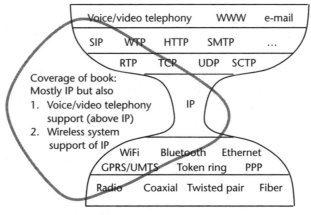

Figure 1.5 The IP hourglass and book coverage.

issues are important and are covered in other texts. For example, network management in IP networks is covered by Stallings [5].

Assumptions on reader background are minimal. I assume a basic understanding of networks, including some knowledge of layering concepts, such as the traditional seven layers of the protocol stack, that are readily available from texts like that of Tanenbaum [6]. This book's preliminary Chapters 2 and 3 may be skimmed or skipped if the material is familiar.

1.6 Summary

We start by painting a picture of a future where people's lives would be enriched by a variety of new, enhanced applications. We focus especially on applications that would be built upon an advanced telecommunications infrastructure, integrating wired and wireless links and internetworked by Internet technology. We discuss the requirements for making this type of vision a reality. From these requirements flow the preview of the book, where we see how the book surveys the technologies that make the vision possible. We then introduce some general themes and principles that are seen throughout the book, and, last, lay out the scope of coverage of the book.

References

[1]   Foster, C., "Totally On," *Stanford Magazine,* May/June 2004, pp. 43–49.

[2]   Saltzer, J., D. Reed, and D. Clark, "End-to-End Arguments in System Design," *ACM Transactions in Computer Systems,* November 1984, pp. 277–288.

[3]   Isenberg, D., "The Dawn of the Stupid Network," *ACM Networker,* February/March 1998, pp. 24–31.

[4]   Deering, S., "Watching the Waist of the Protocol Hourglass," *International Conference on Network Protocols (ICNP),* 1998.

[5]   Stallings, W., *SNMP, SNMPv2, SNMPv3 and RMON 1 and 2,* Reading, MA: Addison Wesley, 1999.

[6]   Tanenbaum, A., *Computer Networks,* 4th ed., Upper Saddle River, NJ: Prentice Hall, 2002.

CHAPTER 2

The Internet

The Internet is a global network of networks of computers. The "network of networks" concept is also known as internetworking and thus is even incorporated into the name Internet. It refers to the interconnection of many different networks, based on multiple technologies (e.g., Ethernet, token ring, and satellite links), into a single, larger network. Why not instead have one global network based on a single technology? Since no single network technology is best for all situations, there are different network technologies in use in different networks. Mandating a single technology for the global network would be very difficult, technologically, since painful compromises and other difficult decisions would have to be made in designing that single technology. Technology aside, the political obstacles would also be formidable; take the 3G wireless systems as an analogy. Despite the efforts of the International Telecommunication Union (ITU), a single unified global 3G wireless system failed to emerge (for more details, see Chapter 3).

Therefore, the internetworking concept arose as an alternative that would maintain the multiple network technologies while interconnecting the disparate networks into a single network. Thus, advantages of a single large network, such as connectivity between any two computers in the network, would be obtained without requiring a single underlying network to be used. This can be seen as an example of convergence in action, where internetworking was the glue that brought the disparate networks into a single, converged network.

In describing the Internet as a global network of networks of computers, we use the term "computers" loosely to include machines with computing power not limited to what one could buy as a "computer" in a store. For example, at the edge of the Internet, the end devices could include large mainframe computers, the ubiquitous personal computers (PCs), laptops, palmtops, personal digital assistants (PDAs), cell phones, and "skinny clients" (cheap, bare-minimum machines that derive their usefulness from being clients of services run on more powerful servers). The Internet infrastructure could include routers (machines that forward packets between other machines) and gateways (machines situated on the boundary of two or more disparate networks, allowing communications to pass between the networks). All these end devices and infrastructure devices are considered "computers" in the loose sense. From a network perspective, the computers that are in the network are known as nodes of the network.

Nodes in the Internet can be divided into *routers* and *hosts*. Generally, a router is a machine that forwards packets; in other words, when it receives packets not destined for it, it does not discard them but sends them to another machine (hopefully moving the packets forward towards the correct destination). Sometimes, the term "router" is used in a more specific sense, to refer to dedicated hardware optimized to perform packet forwarding.[1] In contrast, a host is an end user of the network, such as a PDA or laptop. One way to think of hosts and routers is that for typical Internet applications like Web surfing and e-mail, the traffic originates and ends at hosts, whereas the intermediate nodes traversed by the traffic are routers. In Figure 2.1, for example, the circles, ovals, and PC on the left and right are hosts, whereas the boxes in the middle are routers. A source of possible confusion is that the term "host" can also be used in a different context, for instance, "Web hosting," or used more generally in the sense of a server containing ("hosting") various services for clients. In this book we will use the term "host" only in the classical Internet terminology sense (i.e., hosts are end users of the Internet).

We distinguish the Internet (the one global network) from the technology that makes this possible, called Internet technology. Internet technology includes all the services and protocols used in the Internet. At the core of Internet technology is the Transport Control Protocol/Internet Protocol (TCP/IP) technology that will be discussed in Section 2.3. Since TCP/IP technology works so well and has excellent support in products from numerous vendors, some organizations have TCP/IP-based private networks that they choose to not include in the Internet. These private networks may even span multiple countries on multiple continents and include tens of thousands of nodes. Consequently, the Internet is sometimes also referred to as the "global Internet" to distinguish it from just any other collection of networks that use Internet technology; see Figure 2.2.

As introduced in Chapter 1, historically the voice-centric circuit-switched networks have existed, as well as the datacentric packet-switched networks. IP technology falls in the datacentric packet-switched camp (however, as will be discussed in Chapter 4, efforts are under way to transport voice over IP as well). Since

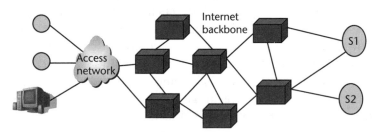

Figure 2.1   Routing problem illustrated.

---

1.   In this book, it should be clear from the context whether we mean "router" in the more general sense or more specific sense. The more general sense should be assumed in cases of possible ambiguity.

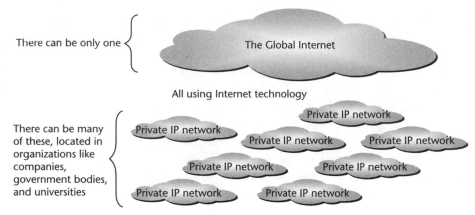

Figure 2.2   Distinguishing the global Internet from other IP networks.

packet-switched networks route data on a packet-by packet basis, an introduction to IP routing will be provided in Section 2.2.

Hundreds of millions of computers, including tens of millions of Web servers, are part of the Internet. These computers are made by hundreds of manufacturers and are configured and managed by thousands of administrators. Without a common understanding of how they communicate with one another, they would not be able to do so, and the Internet would not exist. In fact, a common understanding does exist, on many aspects of the Internet, including how to construct and interpret packet headers and how to decide where to route packets. The understandings on these issues are known as the Internet protocols. We discuss Internet protocols further in Section 2.3.

The protocols that specify how computers communicate with one another, as part of the Internet, are formalized in Internet standards. Internet standards specify the way that components of the Internet should work (see Appendix 2A.1).

2.1   Short History

The Internet began as a series of experiments in internetworking, inspired in part by Kleinrock's seminal work on packet-switching theory, and Licklider's concept of a "galactic" network. In the 1960s, the Advanced Research Projects Agency (ARPA) funded early research on computer networking. Bolt Beranek and Newman (BBN) was contracted to build the initial ARPANET, the network which later became the Internet. The earliest messages on the ARPANET may have been between researchers in University of California at Los Angeles (UCLA) and in Stanford Research Institute (SRI). In those days, TCP/IP had not yet been developed, and the early ARPANET used the Network Control Protocol (NCP) instead.

In the 1970s, pioneers like Kahn and Cerf created the TCP/IP protocols, based on the emerging concept of internetworking. Since ARPA had funded research of various networking technologies, including packet radio and satellite technologies, it became important for the new internetworking protocols to support multiple network technologies. In the early 1980s, ARPANET transitioned to using TCP/IP instead of NCP and became the backbone for the Internet. Later in the 1980s, the NSFNET, a network funded by the National Science Foundation (NSF), also adopted TCP/IP and then became part of the Internet.

More and more computers were connected to the Internet, and more and more networks became part of it. Like a growing snowball, it quickly increased in size. As size increased, so did usefulness, leading to further growth. TCP/IP became the dominant data networking protocol in the world. Factors contributing towards the success of TCP/IP include:

- *Ease of internetworking.* The TCP/IP design includes features like decentralized routing decisions and routers that forward IP packets regardless of underlying link technology, which facilitates internetworking.
- TCP/IP was included in the Berkeley Software Distribution (BSD) implementation of the popular Unix operating system, giving it an advantage with the many computer scientists who used Unix.
- The research funding and gentle guiding hand of DARPA helped the TCP/IP networking technology to mature, and to be widely disseminated to universities and research institutes. Even after the ARPANET itself was decommissioned in 1990, the Internet had grown way beyond a critical mass and continued to survive and grow.
- LAN technology dovetails neatly with TCP/IP, since LANs specify the "subnetwork" transport (i.e., the communications below the network layer of the protocol stack), and TCP/IP the networking over that. The introduction of LAN technology, and Ethernet technology in particular, helped the growth of TCP/IP, coupled with the previous factors (especially since many of the early adopters were the same universities and research institutions using TCP/IP).

2.2 Routing

We previously illustrated the routing problem in Figure 2.1. A typical PC user (represented by the PC icon on the bottom left) wants to access a server (e.g., S1 or S2 on the right), perhaps to browse a Web page. The PC is connected to the Internet through an access network. The access network could be one of many possibilities, including wire-line alternatives like telephone-line modem dial-up, cable modem, or DSL access. It could also be one of the wireless alternatives, such as wireless LAN or GPRS, that will be discussed further in Chapter 3. We note also that the actual global Internet is not as simple as what is pictured in Figure 2.1. First, there are many more nodes in the Internet, so in general the number of *hops* (number of nodes through which packets must traverse) between a PC and a Web page would be larger. Second, the core of the Internet is not monolithic but is comprised of the networks of multiple Internet service providers (ISPs) of different sizes (see Section 2.4.1).

The problem (or challenge, depending on your perspective) of Internet routing is strongly related to packet switching. Because no circuits (or virtual circuits) are established prior to transmission of a packet, IP packets contain all the information needed to handle and route the packet, in the "front" portion of the packet called the header. The header will be discussed in detail in Appendix 2A, but for now, we note that it contains the source and destination IP addresses (the end points for the path the packet will take). Another consequence of the packet-switching architecture is that IP provides *best-effort* packet delivery with no guarantees on packet delivery time, the path the packet will take, or even whether the packet will arrive at its destination. The implications on QoS will be explored in Chapter 7. Meanwhile, in this chapter, we will explain later how TCP provides reliability even over the best-effort underlying packet delivery of IP.

Consider the path an IP packet will take. End points are specified by IP addresses. Between the two end points, the packet goes through routers (recall the difference between hosts and routers, as discussed earlier). However, the path taken between the end points is not necessarily fixed. Instead, a distributed routing mechanism is employed. Each router makes an independent decision, based on the routing protocol it is running. Before we can discuss the routing decisions further, we need to introduce IP addressing.

2.2.1   Addressing

An IP address is a 32-bit number[2] with the following properties:

- All destinations towards which an IP packet can be routed are represented by an IP address, the destination IP address. Similarly, all IP packets have a source IP address;
- Most IP addresses are unique. However, certain blocks of IP addresses are reserved for private usage. These can be used within organizations for internal networking, only on their internal networks. These addresses are never "exposed" to the global Internet (i.e., they are not advertised outside the internal network).
- An IP address identifies an IP connection rather than a machine. This distinction is important in understanding why *multihomed* computers work. The same machine can have multiple IP addresses, one per network interface, for two or more interfaces to different networks. Hence, the machine has multiple home networks. Figure 2.3 shows an example of a PC with two interfaces, with the addresses 200.3.21.5 and 200.31.55.197.

---

2.   The 32-bit addressing is true only for IPv4, the current version of IP. The next version, IPv6, uses 128-bit addresses. IPv6 will be introduced in Chapter 9.

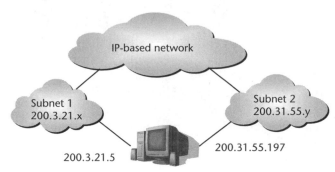

Figure 2.3   Multihoming illustrated.

An IP address is comprised of two parts, a *network address* and a *host address*. If the network address is *n* bits long, the host address is the remaining 32-*n* bits. The network address is always before the host address (see Figure 2.4), so it is also called the *network prefix*. Two addresses with the same network address but different host addresses are considered different connections in the same network (we will soon explain, in Section 2.2.2, how this fact is exploited by the routing algorithms). In the traditional address classification scheme, *n* could be 8, 16, or 24 for class A, class B, and class C addresses, respectively. Clearly, the host address portion has the most room in a class A address, with room for 16,777,216 (2^{24}) different host addresses, whereas there is room for only 65,536 and 256 different addresses per class B and class C network prefix, respectively. Class D addresses are for multicasting and class E addresses are for experimental purposes. In multicasting, packets are sent to groups of recipients (identified by a class D address) rather

Network address	Host address

Class A

Network address 0xxxxxxx	Host address

Class B

Network address 10xxxxxx.xxxxxxxx	Host address

Class C

Network address 110xxxxx.xxxxxxxx.xxxxxxxx	Host address

Class D

Network address 1110xxxx.xxxxxxxx.xxxxxxxx	Host address

Class E

Network address 1111xxxx.xxxxxxxx.xxxxxxxx	Host address

Figure 2.4   Classful addressing.

than to a single recipient (also known as unicasting). Multicast and unicast addresses (class A, B, and C) do not overlap; an address is a class D address if and only if its first four bits are 1110. Similarly, class E addresses, used for experimental purposes only, can be uniquely identified by their first four bits (1111).

The original idea was for a one-to-one correspondence between network address and network. However, the address classification scheme was soon found to be inefficient. It became more likely that, as time went on, a single organization would have numerous networks (e.g., many different LANs). If there had to be a one-to-one correspondence between each LAN and network address, the addresses would be used very inefficiently. Even small LANs with a handful of machines would require at least a class C address, enough for 256 addresses. Therefore, the *subnet* concept was introduced, so the same network prefix could be shared by multiple networks (e.g., multiple LANs). The subnet concept exploits the hierarchical nature of Internet routing by extending the network prefix by a few bits so that each subnet has the same network prefix but different extended prefix. However, this extension is understood only locally, so the rest of the routers on the Internet still route based on the network prefix. Only once the packets reach the subnetted network, do the local routers (in the subnetted network) understand and interpret the extended prefix to be associated with different subnets.

Subnetting helped with one problem: efficient address allocation within an organization, once the organization had a sufficient pool of addresses to work with. But another problem is how to assign addresses to organizations efficiently in the first place. Consider the case of an organization or university wanting to have a 90,000-node network. It would be unable to use class B addressing, since 90,000 exceeds 65,536, and so it would have to use one of the few class A network addresses available. Similarly, a small network needing 1,000 addresses would have to obtain a class B network address. Solutions to this problem of inefficient utilization of the limited 32-bit address space include:

- *Classless addressing* so that organizations do not have to choose between class A, class B, and class C address spaces. This reduces wastage of IP addresses.
- *Private addresses* with *address translation* allows an organization to internally assign to machines addresses that are not globally unique. This allows controlled reuse of IP addresses.
- Unofficial assignment of IP addresses to machines in private networks within an organization, where these IP addresses have not officially been assigned to the organization for its usage.

These three solutions are generally acknowledged to be short-term solutions. In the long run, it is commonly believed that the replacement of IPv4 by IPv6 will be necessary to alleviate the address shortage problem. Chapter 9 provides more details. We now discuss the three short-term solutions in turn.

In classless addressing, the old concepts of class A, class B, and so on are eliminated, so an IP address could have a variable-length network prefix, and not just 8, 16, or 24 bits. Furthermore, the intermediate routers forwarding a packet *do not* know what the network prefix of the packet is! Addresses in this scheme are known as classless addresses or classless interdomain routing (CIDR) addresses, after a Request for Comment (RFC) that described the scheme and suggested how to route packets with such addresses. A hierarchical routing concept has been introduced to work with classless addresses. This will be discussed shortly, but first we introduce some notation. Previously, with classful addressing, the first few bits identify the class, and hence the number of bits in the network address. Clearly, this is no longer true once classless addressing is used. For classless addressing, the network address is often written in the format w.x.y.z/v. For example, if the network address is 192.46.0.0/16, that means the first 16 bits are the network portion of the address, whereas 192.46.0.0/20 means that the first 20 bits are the network portion of the address. In both cases, the network address begins the same way.

A private address, at least officially, is an address in one of the 3 designated blocks for private addresses [1]. These are:

- 10.0.0.0 to 10.25.255.255;
- 172.16.0.0 to 172.31.255.255;
- 192.168.0.0 to 192.168.255.255.

Unlike regular IP addresses on the Internet, private addresses are not made public and they are not globally routable. Instead, we can expect to find the same private addresses used in numerous different networks. There is no conflict, because the scope of usage is only within the organization or group of organizations. Internally, though, each private address should be unique.

This is sufficient for internal usage of private addresses. However, in many cases, organizations have only one or a few global, public IP addresses, and numerous machines that they nevertheless wish to equip with Internet connectivity. Often, these machines are assigned private IP addresses, and a network address translator (NAT) is placed at the boundary of the internal and external networks. The NAT performs address translation between the private addresses and the global addresses. Since there are often more private addresses than global addresses, the NAT needs to keep track of some state information (such as port numbers) to allow it to correctly map incoming traffic to the appropriate private addresses.

Finally, if there are machines within an organization that do *not* need Internet connectivity, but the organization still wishes to use IP to network these machines, arbitrary private addresses can be assigned. These addresses need not be restricted to the three designated blocks of official private addresses, because traffic in the private network never goes to the Internet and so will not result in any conflicts with other machines.

2.2.2 Hierarchical Routing

IP routing uses distributed table lookup. Each router has its own routing table so that whenever a packet arrives at the router, it looks up its routing table to decide where to route the packet. In this section we will explain:

1. Given the routing table, how the router decides where to route the packet;
2. How the routing table is created in the first place, and how entries are inserted and removed.

Entries in the routing table are pairs of either *host addresses* or *network addresses,* and a corresponding outgoing network interface. A host address is, as can be expected, the IP address of a single host (more precisely, it is the address of an IP connection on a single host). A network address is the address of a network, and it is basically an address prefix, where the network address applies to all IP addresses that match the prefix. For example, 192.46.1.26 and 192.46.56.134 both match the prefix 192.46/16, but 192.41.33.1 and 18.244.111.23 both do not match it. Furthermore, the prefix 192.46.1/24 is a more specific match for 192.46.1.26 than 192.46/16, since fewer addresses would match 192.46.1/24. The key idea is that the router tries to make the most specific match for a given destination address. Recall that the router does not know the real network prefix of the IP address. This routing scheme contains the implicit and important assumption that nodes are arranged somewhat hierarchically by address, so that routing by the most specific match works even if the real network prefix of the IP address is longer. If not even a single match can be made, the *default route* is used, if it exists.

How the routing table is created in the first place is either "manually," with *static routes* entered by a system administrator, or using routing protocol software that dynamically updates the routes (hence, these are *dynamic routes*). Popular routing protocols include Routing Information Protocol (RIP) and open shortest path first (OSPF). Actually, I think the term routing protocol is a bit of a misnomer, because these protocols are really protocols about distributing routing information (i.e., routing information protocols). Discussions of these protocols is beyond the scope of this chapter, but the reader may refer to Perlman's book for more details [2].

The choices made in designing the IP routing schemes and protocols have far-reaching implications for all aspects of Internet protocol development. As we will see in Chapter 6, the IP mobility protocols have been constrained in fascinating ways by the existing routing protocols.

2.3 Protocols

Internet communications is designed in a layered framework (the seven layers of the communications "protocol stack" are, from the bottom up, physical, link, network, transport, session, presentation, and application layers. Refer to Tanenbaum's text *Computer Networks* for comprehensive coverage on layering principles and practice [3]. TCP and IP are the two major protocols used in the Internet (and more generally, in all IP-based networks), hence the abbreviation TCP/IP. TCP is a transport-layer protocol and IP is a network-layer protocol. Above TCP/IP, the session, presentation, and application layers are typically not clearly distinguished in IP-based networks. In this chapter, we will introduce IP and TCP. Other protocols relevant to QoS, security, VoIP, and mobility will be discussed later in the book.

2.3.1   IP

The IP is something of a misnomer because it is only one of the protocols that make the Internet work, albeit a foundational component of the network layer of IP-based networks. IP specifies a packet-header format and addressing scheme for IP-based networks, as well as how to forward received packets based on the header information and other criteria. The packet header is a necessary component of packets used in packet-switched connectionless networks. It is a form of overhead and one of the prices paid for using connectionless networks. As part of the IP protocol, all IP packets need to follow the specified format, so that the intermediate nodes and destination node will know how to process the packets. The IP header was carefully designed to contain all the fundamental information needed for correctly processing IP packets in transit.[3] The packet header is typically 20 bytes long, except when IP options are specified. We provide more details on the IP header in Appendix 2A.1. Note that the coverage here is on the present version of IP (i.e., IPv4). The next version of IP, IPv6, has a different header that will be discussed in Chapter 9.

2.3.2   TCP

As mentioned earlier, IP does not guarantee on-time packet delivery, nor does it guarantee whether the packet will arrive at its destination. In some cases, applications require more reliable transport of packets, while in other cases, applications do not require more reliable transport of packets. For example, consider an application for transferring a video file. First suppose that the file is being transferred in order to make an accurate backup of the video file. In this case, reliable delivery is important. Lost packets must be re-sent, even if it takes more time to do so. Next, suppose that the file is being transferred in *streaming* fashion to a recipient who wants to watch the video stream "live." In this case, a few lost packets once in a while are not important, and it is more important to ensure that the stream continues without significant added delay. Reliable delivery (which resends lost packets) is wasteful and unnecessary in this case, since by the time the lost packets are realized as lost, they are not needed. The viewer does not want to wait for retransmitted packets that interrupt the viewing of the video. Notice that for the same task, to transfer a video file, the

---

3.   As the Internet has evolved, it has grown to meet new needs. To provide QoS, security, or mobility enhancements, various techniques are deployed to add functionality without breaking the working of IP in intermediate nodes that may not understand the enhancements. We discuss this further in Section 2.4.4 and in other chapters.

application may or may not require reliable delivery, depending on the intended usage of the file.

With the realization in mind that different applications may or may not require transport, the designers of the Internet have provided a reliable transport protocol, TCP, as well as an unreliable one, user datagram protocol (UDP), either of which can be chosen by applications as the transport protocol to use over IP. When sent over IP, the TCP or UDP packets will be *encapsulated* (carried) in IP packets, in the "data" portion of the IP packet (the "data" portion in Figure 2.5). TCP and UDP have their own headers as well. However, these transport protocols are end-to-end; in other words, the intermediate nodes process only the IP headers and forward the packets without concerning themselves with the TCP or UDP headers, or the rest of the data. At the destination, the TCP or UDP packets are *unencapsulated* from the IP packets in which they are carried, and processed accordingly.

TCP provides reliability in the sense that all packets sent are eventually received (lost packets are retransmitted until received) as well as in-sequence delivery of packets. Reliability is achieved by using TCP acknowledgment (ACK) messages sent from the receiver to the sender. In order to avoid having to wait for the ACK after sending each packet, TCP senders come standard with a *TCP transmission window*. If the TCP transmission window is of size n, that means that up to n packets can be transmitted without the sender needing to received an ACK (i.e., up to n unacknowledged packets can be pending). Choosing the transmission window size involves some tradeoffs. For example, Figure 2.6 illustrates how a transmission window size that is too small can slow down the packet flow (the picture on the left), whereas a

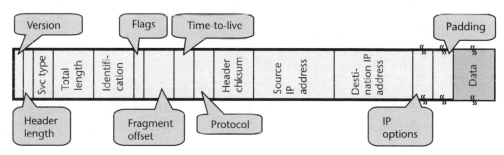

Figure 2.5   IP header.

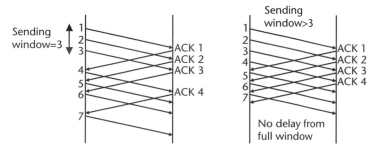

Figure 2.6   Use of TCP transmission window.

larger transmission window size does not slow it down. However, with a larger transmission window size, more packets could be lost in certain situations. TCP is a complex protocol with many other tradeoff and optimization issues. There are even different variations of TCP with different sets of optimizations (these go by names such as TCP-Tahoe and TCP-Reno). More details can be easily found in the literature.

TCP also contains flow control and congestion control mechanisms. However, these were neither designed nor optimized for wireless link characteristics. For example, because of higher error rates on wireless links, more packets may be dropped as they traverse wireless links. However, TCP flow-control mechanisms use dropped packets as an indication of congestion, so they may mistakenly trigger congestion control mechanisms even when there is no congestion or little congestion. Even when packets are not dropped, they may experience significant latency traversing wireless links, resulting in TCP timer time-outs that may trigger congestion-control mechanisms as well. Solutions for such problems are discussed in Balakrishnan [4].

Why was TCP needed? Why was not all this built into IP in the first place? One reason is that reliable delivery is sometimes not necessary (perhaps because the application itself takes care of that), and UDP/IP can be used. A second reason is that the TCP/IP split is a cleaner, more elegant implementation of layering concepts. IP focuses on network layer tasks like routing and packet forwarding, and handles them well. TCP focuses on transport layer tasks like flow control and congestion control, and handles them well.

2.4   Building the Internet

2.4.1   The Internet in Practice

From Figure 2.1, one may get the idea that the Internet consists of nothing but hosts (many of them acting as clients and servers for various services) surrounding a core network of routers that provide the interconnections needed for communications between the hosts. This is a somewhat idealistic picture; in reality, connections to the Internet are typically through commercial entities called Internet service providers (ISPs). Hence, we may envision a more realistic view of the Internet with ISPs interconnected through a "backbone" core network of high capacity routers; see Figure 2.7.

But in practice, who will provide the backbone? In fact, the backbone itself consists of the networks of different providers, as shown in Figure 2.8. Generally, there are regional networks, provided by local ISPs (also known as Tier 3 ISPs), which obtain *transit* connectivity to other parts of the Internet through larger networks provided by larger-scale ISPs (e.g., ISPs with coverage across entire nations, also known as Tier 2 ISPs). These in turn obtain transit connectivity from the few global ISPs (also known as Tier 1 ISPs). Often, an ISP that obtains transit from another ISP pays that ISP for the service roughly proportionately to the traffic volume. In

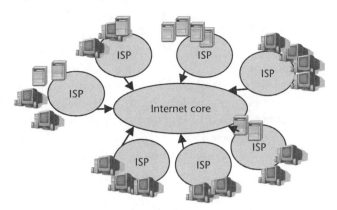

Figure 2.7   A simplified view of the global Internet.

addition to transit relationships between ISPs, there are also *peer* relationships between similar-scale ISPs that wish to reduce their transit costs. In Figure 2.8, transit connectivity is shown using arrows pointing from the lower-tier to higher-tier ISPs, and peer connectivity is shown using dotted lines. Note that although only one Tier 1 ISP is shown in the figure, multiple Tier 1 ISPs exist throughout the world. Since different people administer different "parts" of the Internet, we say that the Internet consists of *administrative domains*.

In this book, we focus mainly on what goes on at the edges of the network, especially when links to hosts could be wireless links. In connecting to the Internet, a wireless network provider (e.g., cellular network or wireless LAN network provider) would face largely the same issues as any other ISP regarding connectivity requirements, and establishing transit and peering relationships.

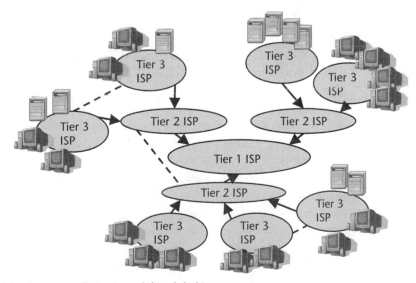

Figure 2.8   A more realistic view of the global Internet.

2.4.2 Design Philosophy

A good paper on the design philosophy of the Internet protocols is Clark's analysis [5]. A discussion of some of the architectural principles and philosophy of the Internet can be found in Bush and Meyer's Request for Comment (RFC) [6]. They highlight the "simplicity principle," and provide an interesting discussion of the comparative merits of circuit switching and packet switching. The principle itself is simple, given how quickly large systems can become complex: "...to be successful we must drive our architectures and designs toward the simplest possible solutions" [6].

In practice, the simplicity principle is seen to be manifested in the way protocols for the Internet are specified. The protocols mostly attempt to solve manageable problems of reasonable size, rather than attempt to achieve too much. Instead of a large, monolithic specification for how the Internet should work, many RFCs have been written to address smaller aspects of how the Internet should work. Thus, the Internet is designed in a modular structure, with protocols designed for simple sub-problems and designed to handle them well; the protocols work together to produce the resultant many-featured Internet. This is arguably a main reason why the Internet has been able to adapt so well to changing requirements (see Section 2.4.4). It is feasible to add new modules (protocols), or make modifications to existing modules (protocols) as necessary. It is interesting to compare with the traditional telephony systems approach, where the systems are well integrated and highly reliable, but perhaps less flexible and adaptable to change.

The success of SIP (see Chapter 5) can be interpreted as a success story for the simplicity principle. SIP started simple and lightweight, a text-based protocol designed with a few basic goals, and not encumbered with lots of extraneous features. It did what it was meant to do well, and left other functions to be provided by other Internet protocols. Gradually it has evolved to add necessary features (as we will see in Chapters 5 and 12), but the SIP community continues to be vigilant to ensure that it does not become bloated with unnecessary features. Thus, SIP has done well against competing protocols for session control of multimedia-over-IP sessions.

2.4.3 Applications

Many people think that the Internet has been around only for about 10 years, not realizing that it began in the 1970s. The reason for this apparent short-sightedness is that the Internet and the World Wide Web (WWW) are synonymous in the minds of many people. Since the WWW emerged around 1994, that year became the starting point for the Internet in people's minds. Actually, before the WWW, the Internet existed, and many universities, government organizations, research labs, and some companies were connected to it. However, there was no one "killer application" that would fuel an explosion in growth of the Internet.

Then the Hypertext Transfer Protocol (HTTP) was created, as were the first Web browsers—and the killer application for the Internet was found. The point is that what ultimately determines the success of technology like the Internet are the applications (what you can do with it). A killer application, like the WWW, can drive the adoption of a technology to newer and greater heights. Electronic mail (e-mail), chat rooms, instant messaging, file transfer, and gaming are other popular applications, which might not be as widespread today if not for the pioneering role of the WWW.

2.4.4   New Requirements and Recent History

In Chapter 1, we stated the principle that systems can be used in ways for which they were not originally designed, and so they should be designed with flexibility in mind. The Internet and its core protocols are a good example of this principle. Commercialization and the growth of the WWW have led to a big transformation of the Internet, both in terms of users and usage. The early users were predominantly researchers, scientists, and university students, a group that forms a narrow segment of society. Usage tended to be for e-mail, file transfer, and remote machine access. Thus, the QoS requirements were minimal—people working in remote logins (e.g., through the telnet application) would be communicating interactively and hence require reasonable latency, whereas people doing file transfers would be more interested in throughput than latency. However, there was not the variety of kinds of traffic that we are seeing today and anticipating for the future. Many applications did not have significant security requirements, if any. For example, machine access might be password protected, but people would often access remote machines through telnet, with their passwords sent in plaintext. With the growing phenomenon of convergence of technologies (and other factors to be discussed in Chapter 4), people have begun to look more closely at transporting multimedia (voice, video, and so forth) over the Internet. The desire of Web sites to provide streaming multimedia content further fuels the research in this area. As a result, control protocols for multimedia session control and streaming control are required in the new Internet. Last but not least, we are observing the convergence of wireless and the Internet, the subject of this book. Because of their differences from wired links, wireless links affect the working of Internet protocols in numerous ways. One major area of impact is mobility—wireless nodes can move. This has implications on such issues as routing, security, and QoS, as will be pointed out in subsequent chapters.

In order to meet the challenges of changing and growing requirements, Internet specifications work has intensified in the last decade. A simple way to measure the rate of activity is in observing the number of RFCs generated in recent years. In the entire previous history of the Internet up to December 1994 (a period of more than two and a half decades), less than half of the total RFCs relating to the Internet were generated. The last RFC published before 1995 was RFC 1735 in December 1994. In the five years till the end of the 1990s, the number jumped up to 2,740 (RFC 2740 was published in December 1999). By May 2004, four and a half years later, the next thousand RFCs had come out, with RFC 3798 appearing in May 2004.

2.5 Further Reading and Summary

The main goal of this chapter is to survey Internet concepts, with emphasis on aspects that relate to the material in the later chapters of this book. Some material is therefore omitted, but the reader can further explore the workings, protocols, and principles of the Internet and TCP/IP in books by Stephens and Comer [7, 8].

In this chapter, basic Internet concepts are introduced, including internetworking, routing, and addressing. Practical aspects of the Internet are discussed, including measures for dealing with the shortage of IPv4 addresses (classless addressing, private addresses, and NATs), and the real structure of the Internet (tiers of ISPs, transit, and peering concepts). A brief history of the Internet is given, and it is noted that much of the history of Internet development is recent history, due to the changing usage of the Internet (e.g., more commercialization) leading to new requirements for QoS, security, mobility, and multimedia transport. We discuss some of the design principles of the Internet and explain how they contribute to the remarkable adaptability of Internet technologies to the new requirements.

References

[1]   Rekhter, Y., et al., "Address Allocation for Private Internets," RFC 1918, February 1996.

[2]   Perlman, R., *Interconnections: Bridges, Routers, Switches and Internetworking Protocols,* 2nd ed., Reading, MA: Addison-Wesley, 1999.

[3]   Tannenbaum, A., *Computer Networks,* 4th ed., Upper Saddle River, NJ: Prentice Hall, 2002.

[4]   Balakrishnan, H., et al., "A Comparison of Mechanisms for Improving TCP Performance over Wireless Links," *IEEE/ACM Transactions on Networking,* December 1997.

[5]   Clark, D., "The Design Philosophy of the DARPA Internet Protocols," *Computer Communication Review,* Vol. 18, No. 4, August 1988, pp. 106–114.

[6]   Bush, R., and D. Meyer, "Some Internet Architectural Guidelines and Philosophy," RFC 3439, December 2002.

[7]   Stephens, W. R., *TCP/IP Illustrated, Volume 1: The Protocols,* Reading, MA: Addison-Wesley, 1994.

[8]   Comer, D., *Interworking with TCP/IP: Principles, Protocols, and Architectures,* 4th ed., Upper Saddle River, NJ: Prentice Hall, 2000.

Appendix 2A   The IP Header

The IP packet header is shown in Figure 2.5. The order of the fields in each IP packet, from "Version" to the optional "Padding," before "Data," is shown from left to right. In the figure, the lengths of the fixed-length fields (the 12 fields from "Version" to "Destination IP Address") are drawn proportionally. The lengths of the other three fields ("IP Options," "Padding," and "Data") are variable. The 4-bit "Version" field specifies the version of the IP protocol used by the creator of the packet (for example, the current version of IP is 4). Hence, routers looking at the "Version" field know how to handle the packet (so they could conceivably handle packets of multiple versions of IP, each in a different way). The 4-bit "Header Length" field contains the length of the header (from "Version" to "Padding") in octets. The 8-bit "Service Type" field may be used to service packets differently based on service classes (more discussion in Chapter 7). The 16-bit "Total Length" field is the total length of the packet in octets (and therefore, the length of the data portion is the difference between this field and the "Header Length" field).

The next three header fields are used to assist in fragmentation and reassembly of IP packets. Given the 16-bit "Total Length" field, the largest IP packet can be 65,535 octets long. However, the underlying link and physical layer technology often can only accommodate shorter packets. Each link and physical layer technology may have its own *maximum transfer unit* (MTU) (i.e., maximum packet size). For example, Ethernet has an MTU of 1,500 octets. As required of a good internetworking design, IP works with many lower-layer technologies. When traffic has to traverse a network segment with a given MTU, IP has the capability to fragment larger packets into smaller pieces that fit the MTU, and to reassemble the original packets after their passing through that network segment. Fragmentation is illustrated in Figure 2A.1. Reassembly involves putting the fragments together.

The 8-bit "Time To Live" field (popularly abbreviated as TTL) indicates how much longer (in number of *hops*) the packet has to live before it must be discarded. Each time a packet is handled by an intermediate machine, the TTL is decreased by one (in the rare event that a packet spends more than one second in an intermediate node, perhaps because of queuing and processing delays, the TTL is supposed to be further decremented to match the number of seconds spent at that node). When the TTL reaches 0, the packet must be discarded instead of being forwarded. The use of the TTL prevents infinite routing loops because the packets will eventually be discarded. Figure 2A.2 shows a routing loop and how the TTL prevents it from becoming infinite. The left subpicture shows the packet going on the most sensible path between "s" (the source) and "d" (the destination). The middle subpicture shows

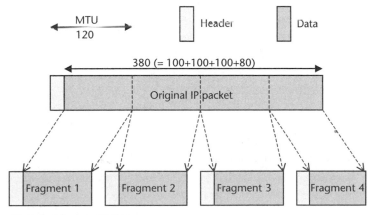

Figure 2A.1   IP packet fragmentation.

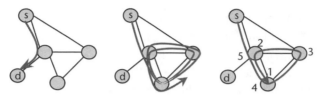

Figure 2A.2   TTL prevents routing loops.

what may happen when a routing loop is in place. The packet could theoretically keep looping forever, but we cut it off after it has completed two loops and is into its third loop. The right subpicture shows how TTL stops infinite routing loops, where in this example an initial TTL of 5 gets decremented to 1. The node at the bottom, before forwarding the packet, decrements the TTL again, and TTL reaches 0, so the packet is discarded and does not loop forever.

The 8-bit "Protocol" field specifies the higher layer protocol (e.g., TCP and UDP) that put the data in the "Data" field of the packet. This allows the contents of IP packets to be processed appropriately. The 16-bit "Header Checksum" field is used for header integrity (see Chapter 8 for more on integrity). The next two fields contain the IP source and destination addresses, respectively. The fields are each 32 bits long, and they contain the original source and ultimate destination even in the case that the packet traverses multiple intermediate nodes. The destination address is an essential input to the routing decision. IP options are usually used only for testing purposes, so most packets do not include them. The optional "Padding" field is used in conjunction with the "IP Options" field in the IP header, to make the length of the IP header an integral multiple of 32 bits.

2A.1   IETF Working Groups

To get an idea where the Internet is headed and how it is developing, technology-wise, it is helpful to see what the IETF working groups are working on.

The IETF standards working groups are divided into eight areas:

- *Applications area:* Deals with applications and higher-layer services, parts of the Internet that make use of the network services (security being an exception that is handled in a different working group). As we will discuss, many aspects of services and applications are best not specified. However, certain applications, like extensions to Internet fax, extensions to File Transfer Protocol (FTP), instant messaging, and presence, are handled in working groups in this area.
- *General area:* Deals with administrative issues related to the workings and procedures of the IETF.
- *Internet area:* Deals with protocols for internetworking, except that routing protocols are handled in the routing area.

- *Operations and management area:* Deals with network operations and management. Network operations includes guidelines for the operation and maintenance of the domain name system (DNS), and for authentication, authorization, and accounting (AAA). Management includes configuration management, performance management, and fault management. Many of the working groups in the operations and management area work on management information bases (MIBs) that can be used for IP network management.

- *Routing area:* Deals with routing, including multicast routing and interdomain routing.

- *Security area:* Deals with security-related issues. We will discuss some of these issues in Chapter 8.

- *Sub-IP area:* Deals with issues beneath the IP layer, such as multiprotocol label switching (MPLS) and asynchronous transfer mode (ATM). A new set of common control and measurement protocols is being used for control of sub-IP technologies like MPLS and ATM that can be used to create "paths."

- *Transport area:* Deals with transport of various kinds of traffic over IP, including SIP, IP telephony traffic, and telephony signaling (sigtrans), as well as other Internet telephony–related protocols like media gateway control. We will discuss some of these protocols and principles in Chapters 4 and 5.

CHAPTER 3
Wireless Networks

The growth of wireless communications in the last two decades of the twentieth century has been astounding. Despite impressive advances in wired (wireline) communications technology in the same period (e.g., in fiber-optic technology), wireless is now a permanent and prominent feature of the telecommunications landscape. People gladly pay to be free to communicate on the go without being tied down to fixed locations.

Long before the telephone was invented, wireless communications existed—in a sense—people shouting to each other over the air between them did not use wires to communicate. However, with the invention of the telephone, electrical signals carried the sound on wires, and thus began the rise of wired telecommunications. Long distance telecommunications was born. While telephony with wirelines greatly increased the range over which telecommunications was possible, it tied users to fixed locations (where the phones were located), and mobility was traded off for range. Mobile telephone systems, when they appeared, regained some of the mobility lost by wireline telephony, without sacrificing range in the end-to-end sense.[1] Since only the segment from mobile phone to base station is wireless and then it plugs into the global phone network, the corresponding party could be very far away, such as on the other side of the globe.

However, wireless links are generally of lower quality and less reliable than wired links. Why? Fundamentally, the wireless communications medium is more difficult than wired communications media. As the signals propagate through the air, they are neither guided by the structure of the wires nor shielded from interference. As a result, the signal energy dissipates rapidly, and interference from other signals can be challenging. In addition, the signals are scattered and attenuated by various objects in the environment like trees and buildings. The same signal arrives at the receiver after traversing different paths, a phenomenon known as multipath that we introduced in Chapter 1. We can only scratch the surface here; for many more details on the wireless communications medium, a classic but still useful reference is Jakes [1].

Thus, in the telephony world, mobile phone users trade off quality for the convenience of mobility. A similar tradeoff is seen in the data communications world.

---

1. In another sense, the range of the wireless link is always restricted; dividing wireless systems by the range of the wireless link is one way to classify them, as will be done later in this chapter.

Ethernet (IEEE 802.3), a wired LAN technology, provides reliable, high data-rate communications. Wireless LAN (WLAN) technology like "wi-fi" (a popular name for WLANs based on IEEE 802.11, a specification in the same family as 802.3) provides mobility to users, but the tradeoff is less quality and less reliability. A general principle emerges: Wireless communications is about tradeoffs first and playing catch-up with wired communications capabilities second. While wireless links continue to improve (with higher data rates and lower error rates), one may observe that they seem to be always a step behind their wired counterparts. For example, by the time WLAN technology has started moving up from 2 Mbps to 11 Mbps, 100-Mbps Ethernet is gaining momentum to replace 10-Mbps Ethernet technology.

3.1 Short History

The discovery of Maxwell's equations and other ground-breaking work on electromagnetism in the nineteenth century, set the foundations for communications using electromagnetic waves. When Hertz discovered in 1886 that electromagnetic waves could propagate not just over wires but also through the air without wires, wireless communications became possible. Hertz also first demonstrated the transmission and reception of electromagnetic waves through the air. The earliest wireless communication systems were the first radio telegraphs demonstrated around the turn of the century. By the early twentieth century, voice (not just telegraphic signals) could be carried by radio. Public radio started taking off [using amplitude modulation (AM)], followed by TV. However, both of these applications of wireless communications are broadcast, one-to-many applications with a large, high-power transmitter that transmits signals to be received by thousands of receivers. Two-way person-to-person long-distance wireless communications were not widely used. Although Al Gross invented the walkie-talkie in 1938, the telephone companies showed little interest in combining wireless and telephony even in the 1950s.

A major development was the creation of the cellular concept in the late 1970s, followed by the deployment of first generation (1G) cellular systems making use of the concept. Previous to the development of cellular, two-way wireless communications systems were used in cities, but these would typically consist of one massive base station that could only support a few expensive user terminals. Multiple simultaneous communications did happen, but they used different frequency channels. Frequency channels are frequency bands used for wireless communications, and separated far apart enough from other frequency channels so that the interference between them is limited. Typically, in a scheme known as frequency division duplex (FDD), a pair of frequency channels is used together for two-way communications, one for base station to terminal (downlink or forward link) and one for terminal to base station (uplink or reverse link). Whenever a pair of frequency channels is in use for communications between a terminal and a base station, that pair of channels cannot be used simultaneously by another terminal throughout the entire city. With the cellular concept, many base stations are used, but the coverage area of each base station is limited. This allows the same pairs of frequency channels used in a cell to be reused in other cells. If all the cells using the same frequency pairs are sufficiently separated from one another, they interfere only minimally with one another. This concept allows *frequency reuse,* tremendously increasing the capacity of the system (the number of users). Figure 3.1 illustrates the cellular concept. On the left is the case where cells are not used. On the right, cells are used, with a frequency reuse factor of three for illustration purposes (i.e., three sets of frequencies are used, for interference reduction; other frequency reuse factors are also possible). The frequency sets are labeled numerically beneath the base station icon in each cell in the figure. Notice that adjacent cells use different sets of frequencies. Also, note that the real wireless propagation environment is not so neat—the hexagons representing coverage areas in this figure are merely a convenient representation. In reality, there would be difficulties placing base stations in such a regular set of positions, and coverage would be of varying qualities in a cell, with poorly covered locations interspersed throughout.

While the cellular concept is a breakthrough in system capacity, it introduces a new challenge—*handoff.* Since users are moving around, and the coverage area of each base station is limited, users inevitably need to switch between base stations, a process known as handoff. As can be seen in Figure 3.1, some of the phones are on the boundaries between cells. These phones would need to hand off between the cells. Handoff is a complex and interesting topic that will be covered further in Chapter 6.

The 1G wireless telecommunications systems are based on analog telephone technology. Voice is carried on analog circuits. Handoff between base stations is network controlled, with all the intelligence and decision-making in the network.

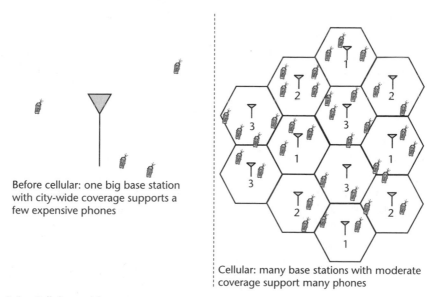

Before cellular: one big base station with city-wide coverage supports a few expensive phones

Cellular: many base stations with moderate coverage support many phones

Figure 3.1   Cellular and frequency reuse concepts.

The electrical circuitry needed to support multiple frequency channels, handoffs, and other such functions was considered state of the art in its time in the late 1970s and early 1980s.

The 2G wireless telecommunications systems are based on digital telephone technology. These systems use digital voice coders (coder/decoders) and digitally coded information streams, as well as the latest digital signal processing. The latest hardware supports the increasing computational requirements. Handoff is also improved, with the mobiles assisting in the decision process. A concept known as *soft handoff* is used in some 2G systems. With soft handoff, instead of simply switching between base stations (a hard handoff), a mobile may be communicating with the system through multiple base stations at the same time during the soft handoff [2]. The mobile eventually decides on one of the base stations, drops the others, and completes the soft handoff. The idea behind soft handoff is to have a smoother handoff process to lessen the service disruptions caused by handoffs.

Interestingly, in going from 1G to 2G, Europe went from many different, country-local systems, to a single unified system, GSM, whereas the United States went[2] from one system, Advanced Mobile Phone System (AMPS), to two different systems, code division multiple access (CDMA) and time division multiple access (TDMA).[3] CDMA, specified in the Interim Standard (IS) 95 (IS-95), and TDMA, specified in IS-54 and more recently in IS-136. We discuss the GSM system in Section 3.2.1.

In 1998, 15 3G radio transmission technology (RTT) proposals from all over the world were received by the ITU. The grand goal was to end up with one RTT for the single global 3G mobile system, whether by selecting one of the proposals or merging a subset of them into one. Unfortunately, this goal was not achieved, due to political differences and deeply vested interests (various parties placed different degrees of importance towards backward compatibility with the existing incompatible 2G systems). Serious attempts were made to reconcile the differences. However, the discussions succeeded only in merging the proposals into five approved RTTs. Of these, the leading proposals were the wideband code division multiple access (WCDMA) and cdma2000 proposals. Both use wideband CDMA in the sense that the bandwidths (5 MHz or more) are wider than the 1.23 MHz used in the 2G CDMA system.

In the world of communications, there are two hemispheres, telephony and data communications (recall the divide between the bell-heads and net-heads as introduced in Chapter 1). As far as wireless communications is concerned, 1G and 2G systems came out of the telephony hemisphere. The response from the data communications hemisphere came in the form of WLANs, which arrived at a time when 2G systems were widely deployed and 3G was being planned. The rapid growth in popularity of WLANs was surprising and caught many people off guard. Companies

---

2. Some would call this a regression, given the benefits of a single unified system.
3. Strictly speaking, CDMA and TDMA should refer to "multiple access" techniques (see Appendix 3A for more details), not just particular systems (IS-95 and IS-54, respectively) that use these techniques.

that had spent billions on spectrum for 3G began getting worried as the WLAN segment of the market grew with astonishing rapidity.

Given the ubiquitous nature of wired LANs in the world of data networking, it is only natural to wonder if LANs could also be built upon wireless links. Although there had been previous attempts to design and sell WLANs, these had been proprietary solutions from individual companies. The solutions were not compatible with one another. Finally, the Institute of Electrical and Electronics Engineering (IEEE) worked to create an open standard that all the vendors could use. The result in 1999 was the 802.11 standard, which is explained in Section 3.2.2.

As the number of computing and communications devices has proliferated in recent years, so has the number of short cables between them. There may be short cables between a PC and a printer or other peripherals. Things would be much neater without lots of messy cabling between devices. One of the original main drivers of Bluetooth was as a cable-replacement technology. This type of idea expanded into the more general concept of wireless personal area network (WPAN). As the name suggests, a WPAN has a shorter range than a WLAN. Apart from Bluetooth, related alternatives include HomeRF, which uses infrared wireless communications. Bluetooth will be discussed further in Section 3.2.3.

In the past few years, a number of new wireless standards have been emerging, including wi-max (IEEE 802.16) and IEEE 802.20. These and other new wireless technologies will be discussed in Chapter 13.

3.2   Types of Wireless Networks

There are different ways to categorize wireless systems. Wireless systems may use licensed or unlicensed spectrum. They have different coverage areas and data rates. Some wireless systems carry voice only, some data only, and some both voice and data. There are also wireless technologies that are less directly relevant to wireless Internet telecommunications (and therefore beyond the scope of this book), including cordless phones, walkie-talkies, and card scanners.

One of the great divides is between licensed and unlicensed spectrum. Usually, if a radio system wants to operate in a particular band of spectrum, the operator must obtain a license from the appropriate governing body. This governing body is typically a regulatory agency such as the Federal Communications Commission (FCC) in the United States, the Ministry of Public Management, Home Affairs, Posts and Telecommunications in Japan, or the Regulatory Authority for Telecommunications and Posts in Germany. The regulatory agency may charge fees (licensing fees) for the use of the desired band and impose certain rules on the use of the band. In some cases, two or more licenses may be granted for the same band(s), where the license owners need to cooperate with one another, within the rules of the regulatory agency, to use the band. Unlicensed operators are forbidden to use the band.

However, a small number of selected frequency bands have been designated as unlicensed bands and follow a different usage model. Any device can transmit and receive wireless signals in the unlicensed bands provided that the *spectrum etiquette* is followed. The operators need not have obtained licenses to use the unlicensed band. What is to prevent utter chaos and a flooding of the unlicensed bands that lead to excessively high interference and poor communications for all users? The answer is the spectrum etiquette—the rules for use of the unlicensed bands are typically much more demanding than the rules for use of the licensed bands. Examples of such rules are that devices use spread-spectrum transmission techniques (to reduce interference to other devices) and that they transmit below certain emission limits. Table 3.1 summarizes the comparison of licensed and unlicensed spectrum.

Unlicensed spectrum comprises the small number of unlicensed bands, and the rest of the available spectrum is licensed spectrum. Most wireless systems, including cellular systems, therefore require licenses. However, popular systems like IEEE 802.11-based WLANs and Bluetooth use unlicensed spectrum.

Coverage areas of wireless systems vary widely. Disregarding satellite systems (yes, they are wireless too, but they are out of scope of this book), the widest areas covered by terrestrial wireless systems are kilometers in diameter. These are the wireless wide area networks (WWAN) like GSM. WLAN, meanwhile, may have ranges on the order of hundreds of meters. For very short-range applications, on the order of meters, we enter the domain of WPAN. We now discuss one leading representative system each for WWAN, WLAN, and WPAN.

3.2.1   Wireless Wide Area Network (WWAN): GSM

The mostly widely deployed and used wireless telephony system in the world today is GSM [3]. GSM provides a full range of services including teleservices speech, fax, and short message service (SMS). A *teleservice* specifies not only the data communications between, but also the terminals. In a layered communications model, this typically means the higher layers services are specified to a substantial degree. This is in contrast to lower-level *bearer services* where only the transport of data between two terminal-modem interfaces is specified. GSM also provides bearer services such as 13-Kbps bearer for voice traffic (which may be used by the speech teleservice) and low-rate data traffic. Additionally, since GSM Phase 2 (GSM was introduced in two major phases, as will be explained in Chapter 11), a large number of supplementary services are also available. These are basic telephony features like call forwarding and call waiting, enhanced in one or more ways for mobile telephony (e.g., call

Table 3.1   Comparison of Licensed and Unlicensed Spectrum

	Licensed Spectrum	*Unlicensed Spectrum*
Use of the frequency bands	Frequency bands are reserved for license holders = less interference	Frequency bands can be used by anybody as long as devices follow spectrum etiquette = potential interference problems
Cost of spectrum	Licensed bands can be very expensive, with costs running up to billions of dollars in highly competitive bands	Unlicensed bands have the great advantage that users do not have to pay license fees to use them

forwarding on busy is different from call forwarding on unavailable), as well as new features like advice of charging, and barring of incoming or outgoing calls according to specific criteria. The GSM lower layers use a TDMA-based approach for multiple access, with Gaussian minimum shift keying (GMSK) modulation.

In this section, we will discuss the following:

* The basic GSM network architecture with basic concepts and terminology;
* Location management, including registration and paging;
* Call setup, and home and roaming cases.

The GSM network architecture was originally very circuit oriented. With the introduction of GPRS and increasing use of 3G, the network is moving towards a more packet-oriented design for data and telephony. Although this book focuses on wireless Internet telecommunications, which is packet switched using IP, we begin with the original basic circuit-switched architecture of the GSM network, for the following reasons:

* Many of the design issues are common to wireless networks in general, rather than just wireless packet switched or wireless circuit switched networks. We will see certain trends emerge that are related to fundamental challenges in wireless networks in general and in handling issues including mobility and security.
* In many cases, solutions for wireless Internet telecommunications are analogous to, or are modified versions of, corresponding wireless circuit-switched telecommunications solutions. Understanding the original GSM design will help the reader appreciate the solutions in the wireless Internet case.

The basic GSM network architecture is shown in Figure 3.2. The user handsets are known as mobile stations (MS), and they communicate with the network over the GSM *air interface* (a set of protocols for communication over GSM wireless links, as will be explained shortly). The network in Figure 3.2 (except for the PSTN cloud on the right) can be thought of representing a public land mobile network (PLMN). Typically, a PLMN is operated by a single operator and restricted to a geographical region like a country. There may be more than one PLMN in a large country, but a subscriber to one operator's service usually would not be able to obtain service from the other PLMNs in the area, unless there is some prior arrangement between the operators. In this case, the service would be handled as a case of *roaming,* which will be discussed shortly.

The GSM standards committees made a wise choice to separate the *subscriber* from the *terminal* (or *equipment*) in the MS. Rather than have each subscriber identified with a mobile terminal, the subscriber is identified with a subscriber identity module (SIM), a card that can be plugged into a terminal. Together, a SIM and a terminal make an MS. Each *subscriber* has a unique international mobile subscriber identity (IMSI) associated with his SIM, whereas each *terminal* has a unique

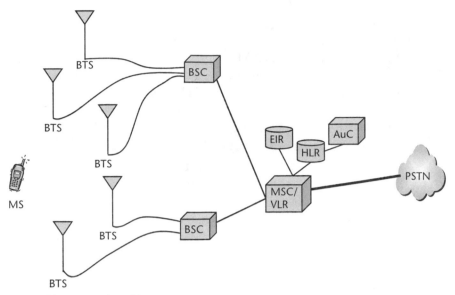

Figure 3.2 GSM network architecture.

international mobile equipment identifier (IMEI). The separation of subscriber from terminal allows one subscriber to use multiple phones and terminals, including perhaps borrowed phones while visiting foreign countries, as long as he brings his SIM along with him and transfers it between phones.

The entry points into the network are known as base transceiver systems (BTS) or, more informally, base stations (BS). The BSs are specialized radio modems and not much more. Rather than putting more intelligence and control functions in the BSs, the GSM designers added another network element, the base station controller (BSC). A BSC with the BSs it controls is referred to as a base subsystem (BSS). On the other side, several BSCs may connect to a mobile switching center (MSC). The MSC is a big machine, handling at any given time the MSs that are using the BSs and BSCs connected to the MSC. It can be thought of as a regular digital switch with added functionality to handle mobile subscribers. For example, the MSC is concerned with radio resource management for the changing set of MSs it handles (unlike a class 4 wireline telephony switch, where the set of subscribers is fixed except for added or canceled subscriptions). Like other digital switches, the MSC is a part of the global signaling system 7 (SS7) network, which uses ISDN protocols.

The added signaling over the SS7 network to handle mobile subscribers is known as the mobile application part (MAP). MAP signaling involves another two elements, the home location register (HLR) and visitor location register (VLR). The HLR is present in every PLMN, and is the database where subscriber information is stored. Even though some operators operate PLMNs in multiple regions or countries, each subscriber signs up for service in one region, so one of the PLMNs is the *home PLMN*. Each of the other PLMNs become the *visited PLMN*, with reference to a particular subscriber, when that subscriber tries to obtain service from it. The capability of a subscriber to obtain service from a visited PLMN is known as *roaming*. The VLR is used to store information on roaming users. The VLR is often integrated with the MSC (we refer to it as an MSC/VLR in such a case). The remaining two network elements in Figure 3.2 are the authentication center (AuC) and equipment identity register (EIR). The AuC is involved in authenticating MSs, a process that will be explained in Chapter 8. The EIR is used to support a theft protection mechanism. It maintains some lists of IMEIs, which can be used to raise a red flag when a terminal reported stolen is used.

3.2.1.1   Location Management

We now look at the concepts of location management, which is part of the solution to roaming (it also supports other goals such as power savings). Location management is a broad concept that includes registration, including storing parts of the MS location information in various databases (HLRs and VLRs) in the network, and the location update and paging procedures. All BSs broadcast their unique cell ID and location area ID (location areas will be introduced in a couple of paragraphs) to assist MSs in location management.

Registration is a procedure in which an MS signals with the GSM network indicating where it is located and that it is "on" and wishes to "attach" to the network. The technical term for registration in GSM is *IMSI attach,* and for deregistration is *IMSI detach*. Recall that the IMSI uniquely identifies a subscriber. If a subscriber is not registered (IMSI detached), the network knows not to bother trying to set up a call to the subscriber. This is a valuable saving of network resources (including avoiding having to page the subscriber, a subject to be discussed shortly). If a subscriber is registered (IMSI attached), the network knows it is worth trying to set up an incoming call to the subscriber. Furthermore, the network acquires knowledge of the location of the user in the process of registration. If a roaming user tries to register in a foreign network, the HLR of the user will be queried and the user authenticated, and then the VLR in the foreign network will obtain part of the subscriber profile and other subscriber-related information (including security related information, as will be elaborated on in Chapter 8) from the HLR of the user. The HLR of the user will also be set up to point to the MSC in the foreign network.

What about location update and paging? The registration described in the preceding paragraph satisfies the necessary location management, as long as the MS does not move after it registers. However, we cannot make that assumption. The MS must be allowed to move, and the network must be updated when it does move. One possibility is to inform the network whenever the MS decides to use a different base station to access the network. When there is active communications going on, this is called handoff to a different base station and is clearly the right thing to do. When active communications does not exist and the MS is in *idle state*, the MS using a different base station is said to be *camping*. It is not immediately clear if informing the network is the right thing to do whenever the MS switches from camping on one base station to camping on another. In fact, the idea behind idle state is to save power while the MS is idle. Hence, in idle state, the MS only updates it location with the network when it crosses boundaries between *location areas*. Location areas are groups of multiple base stations, and the MS knows when it has changed location areas by listening to the base station broadcasts. The use of location areas saves power by not requiring the MS to send location updates whenever it crosses cell boundaries. However, the network now only knows the location of the MS to the precision of a location area. Hence, when an incoming call arrives, it needs to search for the MS within the entire location area. This procedure is called paging. There is an interesting tradeoff between resource utilization for location updates and for paging. The larger the location areas, the more power savings for location updates but the more resource utilization for paging, while the smaller the location areas, the less power savings for location updates but the less resource utilization for paging.

3.2.1.2 Call Setup Signaling

To illustrate how roaming is supported in call setup signaling, we first explain how a subscriber would make or receive calls when at home (in the home PLMN). We can then point out the differences in the roaming case. We use *call flow* diagrams to show the signaling. These diagrams show the relevant network elements at the top, followed by messages sent between them. The vertical lines are to indicate the source and destination of the messages. If the tail of an arrow is touching a vertical line beneath a network element, that element is the source of that message, whereas if the head of an arrow is touching a vertical line beneath a network element, that element is the destination of that message. The sequence of the messages is starting from the top, so the horizontal arrows are arranged in time sequence, with the later messages further down the diagram than earlier messages. In all cases, we designate the calling party as the *caller* and the called party as the *callee*.

When a subscriber in her home network dials a callee, the MSC in the caller's home PLMN analyzes the digits and routes it to the appropriate destination, as shown in Figure 3.3 (you may want to look ahead to the Appendix 5A.1 in Chapter 5 for a brief introduction to signaling in the phone network). The destination may be in the PSTN, another PLMN, or the same PLMN. The signaling between MSC and PSTN switches is very similar to that between digital switches in the wireline PSTN. Except for the two MAP messages (messages 2 and 3), the signaling is basically the ISDN signaling used in the modern digital PSTN (more details in Chapter 5). If the callee is also a mobile subscriber, then the destination side of the signaling will be as described next.

When another party (whether wireline or mobile) dials a mobile subscriber, the call setup signaling gets routed to the home PLMN of the subscriber. It must be routed to the home PLMN even if the subscriber is roaming, because other networks do not know whether the subscriber is at home or roaming. In a PLMN with multiple MSCs, one of them may be configured as the *gateway MSC*. The gateway MSC in a PLMN is the entry point into the PLMN for calls destined to one of its subscribers. The gateway MSC has the responsibility to query the HLR for the subscriber profile

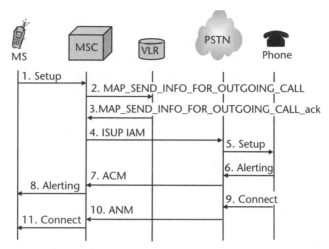

Figure 3.3 Call initiation from MS at home.

and location information. If the subscriber is in her home PLMN, the gateway MSC just has to route the call signaling to the appropriate MSC serving the subscriber.

What about when the MS is roaming? When the MS tries to set up a call, the network that is roamed to will assist with authentication and call setup, based on the protocols and information exchanged earlier during IMSI attach. Otherwise, there is little difference from the case when the MS is at home. However, the situation becomes more interesting when the MS is receiving a call while roaming. In this case, illustrated in Figure 3.4,[4] the call setup reaches the gateway MSC, as when the MS is at home. The gateway MSC will query the HLR, and will find that the MS is

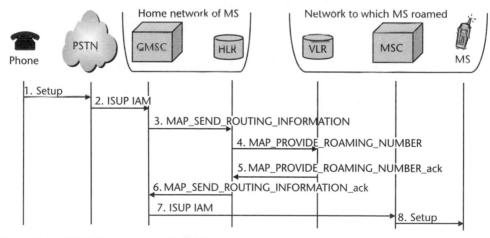

Figure 3.4 Call delivery to roaming MS.

4. Note that we do not show the rest of the signaling after the initial setup reaches the MS, as it is similar to what happens in Figure 3.3.

roaming, as well as in which network it is roaming. These pieces of information are stored as part of registration and updated by location updates; otherwise, if the MS has not registered in the foreign network, the network will be unable to locate the MS. Call setup continues with the next call leg between the gateway MSC and the appropriate MSC in the foreign network. That MSC will route the call towards the appropriate BSC and eventually to the MS. Note that this kind of call setup could result in the voice circuits being much longer than necessary. For example, if the caller and callee are both in one PLMN, but the callee is merely roaming in that PLMN and if the callee's home PLMN is the other side of the globe, the call will still be routed back to the callee's home PLMN, causing long and unnecessary call legs.

3.2.2 Wireless Local Area Network (WLAN): 802.11

IEEE 802.11 [4] is part of the 802 group of standards. The 802 group of standards deals with LANs. Perhaps the most famous member of this group is 802.3, which is basically the Ethernet Medium Access Control (MAC). 802.11 specifies a wireless physical layer (there are three possibilities) and a single MAC layer that works with all three physical layers. Like 802.3, or 802.5 (token ring), 802.11 is supposed to be used with an 802.2 logical link control (LLC), as shown in Figure 3.5. The LLC and MAC are sublayers of the link layer, and together they make up the link layer. A very important difference between 802.11 and GSM is that GSM is a *vertically integrated* system that provides a full range of teleservices and supplementary services, whereas 802.11 provides only relatively low-level physical layer and MAC bearer services.

Machines in an 802.11 WLAN are known as stations. A group of stations that can (and do) communicate directly with one another (because they are in close enough proximity) is known as a base service set (BSS). The BSS is the basic building block of 802.11 WLANs. As shown in Figure 3.6, 802.11 WLANs can be used in either of two basic configurations: ad hoc mode and infrastructure mode. In ad hoc mode, the BSS is an independent BSS (IBSS) that formed in an ad hoc fashion. Infrastructure mode, on the other hand, combines two or more BSSs into an extended service set (ESS) that includes a distribution system (DS), which distributes packets between all the BSSs in the ESS. More precisely, each of the BSSs in the ESS accesses the DS through what is called an access point (AP). Each AP is a regular station with enhanced capabilities. The APs, DS, and stations using the APs together

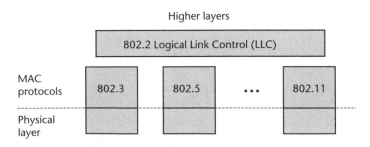

Figure 3.5 802.11 among other IEEE 802 family members.

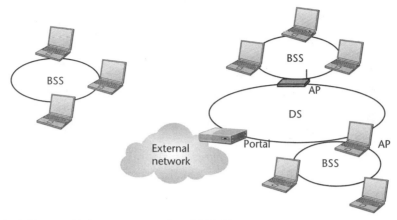

Figure 3.6 Ad hoc and infrastructure modes of 802.11

make up the ESS. In many cases, a functional element known as a portal acts as a gateway to external networks.

An important principle of the 802.11 specifications is that the WLAN should look like an ordinary 802-style LAN to the upper layers. For example, no network-layer IP routing should be needed to send packets from any station to any other station on the same LAN. This applies even for the ESS. The most challenging case is where the DS itself uses 802.11, and where one station is associated with one AP and the other station is associated with a different AP, so packets would have to be sent through the wireless DS. This case is shown in Figure 3.7, where selected header fields are shown with each segment, to illustrate the use of the address fields and two critical flags. The 802.11 header can take not just two MAC addresses (source and destination, as is the case for Ethernet) but up to four. In the case in question, all four address fields will be used when the packet is traversing the DS between stations in different BSSs in the same ESS. The four addresses required are the original source and destination, and the addresses of the two APs involved. The address fields, and the "To DS" and "From DS" flags, are the information needed to allow subnetwork routing to occur through both the APs towards the destination MAC address. The sequence of the addresses in the address fields can be found in the specifications.

What happens when two or more stations are transmitting at the same time? Since the medium (air) is common, there may be interference between the transmissions, resulting in poor performance. This is known as the medium access problem. This problem is not encountered only in wireless, because the Ethernet (IEEE 802.3) MAC protocol shares the Ethernet wired medium using carrier sense medium access with collision detection (CSMA/CD). The medium is sensed before transmission, and transmission happens only if the medium is sensed to be free. However, because of lags between the start of transmission and when another node can begin to sense the transmission, the other node might sense a free channel and begin transmission. In this case, a collision would result but could be detected (hence the "CD" part of

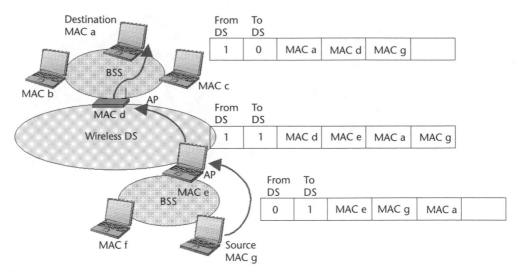

Figure 3.7   Example of use of 802.11 address fields.

CSMA/CD). In the case of a wireless medium, though, there are additional challenges. Although in a wired LAN, every station on a LAN (or LAN segment to be precise[5]) can hear the transmissions from every other station on the same LAN, this is not necessarily true in the wireless case.

Two peculiar situations are illustrated in Figures 3.8 and 3.9. The *hidden terminal* problem occurs when a station (station A in Figure 3.8) wants to transmit to another station (station B) but is unable to hear another terminal (station C) transmitting to station B. Hence, carrier sensing prior to transmitting does not reveal the "hidden terminal" (station C), and station A goes ahead and transmits, resulting in a collision at station B. In contrast, the *exposed terminal* problem occurs when station A is transmitting to station B just as station C wants to transmit to station D.

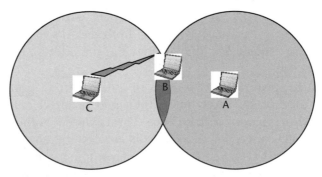

Figure 3.8   The hidden terminal problem.

---

5.   There may be cases where LANs are segmented by smart bridges that may not blindly forward traffic from one side to the other side.

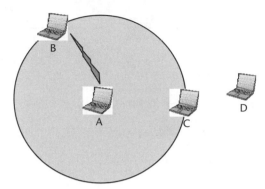

Figure 3.9 The exposed terminal problem.

Based on the topology of the stations, station C would hear station A as it is performing carrier sensing prior to transmission. As a result, station C will not transmit to station D, even though the transmission could have succeeded. Station C in this case is the exposed terminal. 802.11 uses carrier sense medium access with collision avoidance (CSMA/CA), with a request to send, clear to send (RTS/CTS) scheme, which handles the hidden terminal and exposed terminal problems, as will be discussed in Chapter 7.

There are three options for the physical layer of 802.11. Two of these are in the 2.4-GHz unlicensed band, and the other is in the infrared (IR) band (also unlicensed). The 2.4-GHz physical layer options are a physical layer based on direct sequence spread spectrum (DSSS) and a physical layer based on frequency hopping spread spectrum (FHSS). Note that the spectrum etiquette for the 2.4-GHz band stipulates the use of spread-spectrum technology, and these two options satisfy the spectrum etiquette. Although the IS-95 CDMA system also uses DSSS technology, one difference is that IS-95 uses DSSS for medium access (multiple access) as well, whereas 802.11 only uses it for noise and interference rejection. 802.11 uses CSMA/CA for its medium access. One reason for this difference is that IS-95 is a more centrally controlled system, whereas 802.11 is more distributed. Therefore IS-95 can and does coordinate the use of spreading codes.

Since the publication of 802.11, the IEEE has continued to publish enhancements to it. Of note are QoS and security enhancements for 802.11 (in 802.11e and 802.11i, respectively). These are necessary in order for 802.11 to be a serious contender as the access technology of choice for wireless Internet telephony. The standards 802.11e and 802.11i will be discussed in more detail in Chapters 7 and 8, respectively. Table 3.2 lists members of the 802.11 family of standards.

3.2.3 Wireless Personal Area Network (WPAN): Bluetooth

Bluetooth was named after Harold Blatand, a Danish Viking king (Blatand is equivalent to Bluetooth) [5]. Blatand united the Danes, just as Bluetooth was designed to be a ubiquitous short-range radio link that works with all kinds of

Table 3.2 Members of the 802.11 Family of Standards

Specification	Summary
802.11a	A high-rate physical layer (up to 54 Mbps) in the 5-GHz unlicensed bands
802.11b	A high-rate physical layer (up to 11 Mbps) in the 2.4-GHz unlicensed bands
802.11e	Quality-of-service (QoS) enhancements for 802.11
802.11f	Inter-Access Point Protocol (IAPP) for enhanced signaling (including mobility signaling) between access points
802.11g	A suite of high-rate physical layers (up to 54 Mbps) in the 2.4-GHz unlicensed bands
802.11i	Security enhancements for 802.11
802.11n	The latest physical layer under development to deliver the highest data rates yet

devices. The alternative, with dozens or hundreds of different radio protocols for different short-range applications, is roughly analogous to Denmark without the unifying efforts of Blatand. For example, Bluetooth radio links can be used to replace cables, for instance, between a PC and peripherals or between a cell phone and headset, as shown in Figure 3.10. Bluetooth radio links operate in the 2.4-GHz unlicensed band (ISM band), using FHSS. The range is very short (about 10 meters), and the data rates are moderate (up to 723 Kbps).

Networking using Bluetooth is based on the Bluetooth-specific concept of *scatternets* built upon *piconets*. A piconet is a collection of Bluetooth devices that are all synchronized to one of the devices in the piconet, the *master*. There is only one master per piconet, and one to six *slaves*. Or, more precisely, up to six active slaves are possible, but there can be other inactive slaves in a "parked" state that are also synchronized to the master. The slaves are all synchronized to the master in the sense of being synchronized to its clock and hopping sequence. Consequently, there is a Bluetooth radio link between each slave and the master.

A piconet is an ad hoc collection of Bluetooth devices in the sense that there is no need to preconfigure the devices in a piconet to be part of that piconet. In fact, even being a master or being a slave is just a role that a node takes on in the formation of a piconet. For example, a node in one piconet could be the master, and then after the

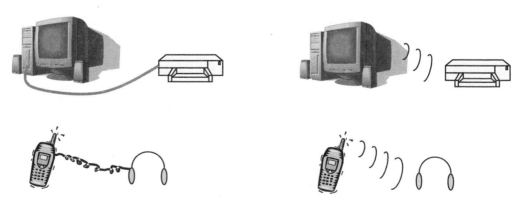

Figure 3.10 Bluetooth as a cable replacement technology.

piconet dissolves (for example, after the nodes in the piconet have moved out of range), it could join another piconet as a slave. The dynamic life of a piconet is simple: It begins when two Bluetooth devices come within range and decide (how this decision is made we will briefly touch upon later) that they wish to establish a Bluetooth link. This decision involves a standard Bluetooth procedure. Thus a piconet is born, with one of the devices as master and the other as slave. Subsequently, other devices wishing to join the piconet do so as slaves, up to the limit of six slaves per piconet. At any time, any of the devices can leave the piconet, up to the point there is only one slave left and it leaves, whereupon the piconet ceases to exist.

Can a master of a piconet be a master of another piconet? Can a slave of a piconet be a slave of another piconet (i.e., can a Bluetooth device have two masters)? Can a master of a piconet be a slave of another piconet? The answers are no, yes, and yes. Suppose a device is master of one piconet. The slaves in that piconet are synchronized to its clock and hopping frequency. If the master were to become the master of another piconet, its slaves would also be synchronized to its clock and hopping frequency. In this case, there is no difference between the two piconets, and they are defined as one piconet. So a device cannot be master of two or more piconets at the same time. However, a slave of one piconet can be a slave in another piconet, on a time-sharing basis. Similarly, even a master of one piconet can be a slave in another piconet at the same time, on a time-sharing basis. Thus, it is possible to form a network with an arbitrarily large number of nodes (not restricted to seven as for a piconet), where a path can be traced from each node to every other node over Bluetooth links. This kind of network, as shown in Figure 3.11, is called a *scatternet*.

Unlike GSM, which was designed originally for voice, or 802.11 WLAN, which was designed originally for data, Bluetooth was designed from the ground up with both data and voice in mind. It supports asynchronous connectionless (ACL) links, as well as synchronous connection-oriented (SCO) links. ACL links provide connectionless transport for packet-switched data traffic, whereas SCO links provide connection-oriented transport for circuit-switched voice and video traffic. A Bluetooth device can use both types of links simultaneously.

Bluetooth features service-discovery mechanisms that allow Bluetooth devices to discover what services are supported by other Bluetooth devices. The Service

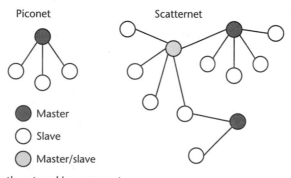

Figure 3.11 Bluetooth networking concepts.

Discovery Protocol (SDP) operates in a client/server fashion. For the purposes of service discovery, a Bluetooth node that wants to discover what services are provided by another Bluetooth node acts as an SDP client. A Bluetooth node that wants to inform other Bluetooth nodes about the services it offers acts as an SDP server. Any node can be both a client and a server simultaneously. A server can furnish information about services provided by multiple applications. SDP clients make requests and SDP servers respond, allowing browsing of the available services as well as searching for services. SDP is informational only; it does not provide a way of accessing the services, of controlling access to the services, or advertising the services. Furthermore, it does not specify how to choose between competing services or how to bill for services.

With its short-range wireless link and ad hoc formation of scatternets, Bluetooth is suitable for a wide range of applications, such as cable replacement for PC accessories, short-range access to a LAN, and cordless headset for a phone. In keeping with the vision of ad hoc networking of Bluetooth devices, while any Bluetooth device from any manufacturer can work with any other Bluetooth device from another manufacturer, certain guidelines are required for the higher layers in addition to the common Bluetooth lower layers. These guidelines are provided in frameworks called usage profiles. Bluetooth specifies usage profiles for services including cordless telephony, intercom, serial cable replacement, headset, dial-up networking, fax, and file transfer.

3.3 Summary

We switch gears coming to this chapter from the last chapter, going from the Internet to wireless technologies. The wireless communications medium is difficult, so wireless links are generally less reliable and of lower quality than wired links, but wireless provides features, such as mobility, that are very attractive. A brief history of wireless is provided, where we trace the development of cellular mobile systems in particular, including the cellular concept, frequency reuse, and handoffs. We also survey the characteristics of 1G, 2G, and 3G systems. However, while cellular mobile systems are a product of the telephony world, other wireless systems (such as WLANs) are a product of the data-networking world. We introduce these systems as well, and consider three examples of wireless systems: GSM, 802.11 WLAN, and Bluetooth. These are examples, respectively, of a wide-area network system, a local-area network system, and a personal-area network system.

References

[1] Jakes, W. C., *Microwave Mobile Communications,* 2nd ed., New York: Wiley-IEEE Press, 1994.

[2] Wong, K. D., and T. J. Lim, "Soft Handoffs for CDMA Mobile Systems," *IEEE Personal Communications Magazine,* December 1997.

[3]   Mouly, M., and M.-B. Pautet, *The GSM System for Mobile Communications*, Palaiseau, France: Cell & Sys, 1992.

[4]   ANSI/IEEE Std. 802.11, "Part 11: Wireless LAN Medium Access Control (MAC) and Physical Layer (PHY) Specifications," 1999.

[5]   Bluetooth SIG, "Specification of the Bluetooth System: Wireless Connections Made Easy," Version 1.1, June 2003.

[6]   Carlson, A. B., *Introduction to Communication Systems*, 3rd ed., New York: McGraw-Hill, 1986.

[7]   Proakis, J. G., *Digital Communications*, 4th ed., New York: McGraw-Hill, 2000.

[8]   Rappaport, T., *Wireless Communications: Principles and Practice*, 2nd ed., Upper Saddle River, NJ: Prentice Hall, 2001.

Appendix 3A   Brief Introduction to Aspects of the Wireless Physical Layer

We here present a very brief introduction to selected terminology and aspects of the wireless physical layer. For more information on physical-layer details, the reader is encouraged to refer to textbooks like Carlson's for a general introduction to communication systems, Proakis' for an introduction to digital communications, and Rappaport's for an introduction to wireless communications [6–8].

There are several concepts of bandwidth that may confuse readers not familiar with wireless terminology. The *information bandwidth* of a digital signal is the bandwidth of the signal in the frequency domain, at baseband (i.e., not modulated by a carrier), as shown in the top left picture in Figure 3A.1. For example, the traditional telephone signals on telephone wires had a bandwidth of 3 kHz. By fundamental information theoretic principles, the information bandwidth is proportional to the data rate of digital signals; hence, a high bandwidth signal often means a high data-rate signal. In digital communications, data is encoded in *symbols* that are

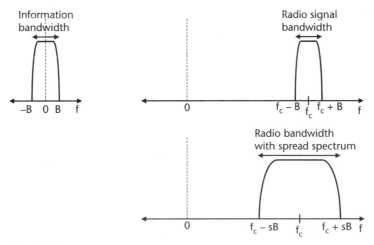

Figure 3A.1   Bandwidth concepts.

transmitted at a certain rate (the symbol rate). In wireless communications, multi-path tends to cause more problems for signals with higher symbol rates.

Wireless signals are normally modulated by a sinusoidal carrier to a carrier frequency. The *radio bandwidth* is the bandwidth of the resulting signal, measured around the carrier frequency. As seen in the top right picture in Figure 3A.1, the radio bandwidth is normally about the same as the information bandwidth, in traditional modulation schemes. However, in a spread spectrum case (as with CDMA), shown in the lower right picture, the radio bandwidth is wider than the information bandwidth, by a factor known as the *spreading factor* (3 in the example, but usually larger than 3). Spreading can be achieved in many ways, but in CDMA cellular systems or the 802.11 DSSS physical layer, it is by "multiplying" the information bit stream with a high-rate code. The code is chosen with good mathematical properties, also known as spread-spectrum properties. For example, in a CDMA system, this allows different users to transmit and receive at the same time with little interference to one another. An alternative is chosen in GSM, which uses a more traditional GMSK modulation without the spread-spectrum properties. In GSM, users are separated by time slots, and hence the term TDMA.

CHAPTER 4

Multimedia over IP

The idea of multimedia communications is very appealing—text, graphics, sound, and video all combined to enrich the ways in which we communicate and to transform communications from a one-dimensional to a multidimensional experience. Many believe that multimedia content will drive the continued growth of the Internet.

Traditionally, voice has been carried on circuit-switched lines. During the past decade, work on VoIP has grown, for a variety of reasons that will be discussed in Section 4.1. More recently, interest in multimedia over IP is rapidly rising. With increasing availability of high-bandwidth links and advances in QoS technologies, a future is often envisaged in which multimedia telecommunications carried over IP networks becomes an ordinary way of communicating.

Our concept of multimedia over IP includes the more familiar concept of VoIP as a special case. Multimedia over IP, as compared to a single media over IP, adds one other complication—the various media streams need to be synchronized. However, whether the subject is a single medium like voice or multimedia, we are dealing with real-time traffic.[1] Real-time traffic has stringent requirements on the playback of the voice and video at the receiving side. Even if a stream is broken into segments (a requirement for transmission in packet form, as is the case for multimedia over IP), the playing of the voice or video needs to be "smooth." In other words, the time relationships between when the segments are played needs to match the time relationships between when the segments were recorded on the sending side. Therefore, variance in the time sequencing of consecutive segments of the voice or video stream (variance, that is, from the time sequencing on the recording side) can be little tolerated.

Furthermore, real-time traffic can be divided into *conversational* and *streaming* traffic [1]. Conversational traffic, such as voice or video telephony, involves an interactive exchange of streaming data. Humans want to perceive the voice and/or video arriving instantaneously or almost instantaneously in order to carry on a normal conversation. Therefore, the end-to-end delay from recording the voice and/or video on

---

1.　Of course, an exception is the case that multimedia content is downloaded for later playback, an example being the transfer of files containing multimedia content over peer-to-peer file sharing networks, and another example being the using of FTP to transfer files, some of which may happen to have multimedia content. For the purposes of this book, however, we are more interested in the cases where real-time playback of multimedia is in effect.

one side to playing it on the other side(s) must be very low. (Studies have one-way delays of less than 150 ms as best, whereas up to even 400 ms may be tolerable in certain conditions, such as when people are physically across the globe and psychologically prepared for small delays [2].) On the other hand, some kinds of streaming real-time multimedia, such as streaming movies from a video server, do not have as tight delay constraints as conversational traffic. Since the traffic is mostly one way, some delay can be tolerated before the multimedia traffic starts playing. However, once it does begin, the usual constraints for real-time traffic apply. Since our focus in this book is on conversational traffic, we will not discuss protocols like Real-Time Streaming Protocol (RTSP), which are applicable for streaming multimedia. Note that RTSP makes use of Real-Time Protocol (RTP) for stream transport. We do discuss RTP in this book, but for RTSP you may refer to the RFC for more information [3].

In Chapter 2, we make a distinction between the Internet and other IP-based networks. Similarly, here we make a distinction between voice over the Internet and VoIP. The global Internet is an existing network of networks that will provide only best-effort packet delivery for the foreseeable future. As such, it cannot provide high-quality transport for VoIP services, much less video over IP services. VoIP, on the other hand, refers more generally to VoIP-based networks. This includes private IP-based networks that can be engineered with the desirable QoS because they are controlled by one organization or a small group of organizations that can agree on QoS controls. With voice over the Internet, each organization only has control over a small piece of the entire network. It is virtually impossible to get every organization using the Internet to agree on a desired QoS solution and to cooperate in implementing it.

4.1 Motivation

We first present a series of possible reasons for development of multimedia over IP and VoIP, and then discuss which of these are primary motivations, and which are secondary.

4.1.1 Efficient Digital Voice Coding

The public switched telephone network (PSTN) previously used only analog transmission of voice. However, the circuits between switches in the PSTN have by now mostly been replaced with digital circuits, where voice is encoded at 64 Kbps using pulse code modulation (PCM). Most lines to homes and offices are still analog lines, although some are also digital, such as integrated services digital network (ISDN) lines. Of course, 64 Kbps is not a magic number, and digital voice coding schemes have been created with other rates as well. For example, in GSM, a 13-Kbps voice coder is used. Interestingly, in GSM, despite the highly efficient 13-Kbps encoding used, there is a transcoder rate adaptation unit (TRAU) in the radio access network to convert between this 13/16-Kbps coding and regular 64-Kbps PCM, because the circuits between the MSC and other switches use standard 64-Kbps PCM. Transporting multimedia over IP could allow the benefits of efficient voice coding schemes to be more fully realized.

4.1.2 Support for Different Levels of Quality and Different Media

With the PSTN, there is no room for different levels of voice quality. It is not a trivial matter to add video telephony due to the design of the circuits that support 64-Kbps PCM voice signals. Higher-rate voice coding schemes could not be easily supported. Even if a better voice coding scheme to "fit" in the 64-Kbps circuits existed, it would cause either wasted bandwidth (it might not need all the 64 Kbps) or other problems in conversion between the A-law PCM used in some parts of the world and the μ-law PCM used elsewhere. (With μ law PCM, switches may assume incorrectly that signals are encoded in a certain way, although they may not even use PCM.) Besides, the PSTN does not accommodate negotiating voice coders, in contrast to SIP in the VoIP case (as will be seen in Chapter 5).

4.1.3 Network and Service Integration

In the past, voice telephony was the main form of network traffic. People might have used voice modems (for data, using the telephone lines for transmissions) to do data communications. Although not very efficient, that was acceptable in the past. As data networking, and in particular, IP networking, has grown, there is less and less need for the PSTN to access networks like the Internet, as various broadband alternatives have emerged. Looking ahead, the volume of data traffic will dwarf the volume of PSTN traffic.

Rather than maintaining parallel PSTN and data networks, might it make sense to integrate them? Given that IP packet-switching technology works very well for general data traffic, could we integrate more "connection-oriented" voice and video traffic over IP as well? With integration, tremendous cost savings could be enjoyed. This is the benefit of network integration. Meanwhile, it might be easier to provide integrated services like unified messaging, or Web and telephony integration, using IP rather than a separate network. Innovative integrated services could be created in such an environment. This is the benefit of service integration.

4.1.4 Statistical Multiplexing

Consider the traditional phone system and the connection between a residential phone and the central office. Lines from several homes in a neighborhood may all be connected to a pole and groups of lines are spliced (consolidated) into larger and larger bundles moving towards the central office. This system is analogous to a river, where the river mouth corresponds to the connection to the central office and the sources of water correspond to the phones. The small streams consolidate into larger streams that flow into the main river as tributaries. One major difference is that in the case of the phone system, not every phone is in use most of the time. At some stages of consolidation, especially closer to the central office, there is not enough capacity to handle all or most of the phones being used *simultaneously*. Hence, it is possible that if many phones are in use, and a call is to be delivered to, or originated from, yet another phone in the same region, that call cannot be completed. Such an uncompleted call is said to be *blocked*.

Clearly, blocked calls are annoying to their victims, so why is not more capacity provided? Suppose that the capacity needed to completely prevent blocking is called full-connection capacity. In this case, there is enough capacity to handle all the phones in the region in use simultaneously. Making some simple assumptions about the duration of phone calls and about the interval of idleness between phone calls, the capacity can be significantly decreased from the full-connection capacity, while maintaining a low probability of call blocking (e.g., 10^{-5}). This is called statistical multiplexing, or the multiplexing of shared resources based on the statistical reasoning that blocking is rare. VoIP arguably provides much better statistical multiplexing than circuit-switched voice sessions, because for circuit-switched sessions, the circuits are tied up even when neither party is speaking, whereas with VoIP, packets are sent only when there is speech to encode and send.

4.1.5 Assessment

Network and service integration is the primary motivation for investment in voice and multimedia over IP. The PSTN is very good for voice, but it is a stand-alone and monolithic network, and mostly a one-medium network as video telephony has not taken off. As the volume of data traffic eclipses and then dwarfs the volume of PSTN traffic, it makes much less sense to maintain two networks just for historical reasons. Integrating the networks will result in cost savings in the end. Also, service integration leads to all kinds of exciting possibilities. As the second important motivation behind network and service integration, I would rank support for different levels of quality and different media. This would fit in nicely with the new service possibilities with service integration. Some bandwidth efficiency may be achieved using more efficient voice coding schemes, as well as through statistical multiplexing, but these efficiencies would be less significant in the future.

Despite the real and perceived benefits of multimedia over IP, QoS and transport challenges remain. The fact remains that the circuit-switched PSTN provides good quality voice using dedicated circuits. With multimedia over IP, the best-effort nature of IP packet delivery is a real concern. If it turns out that acceptable-quality voice and video cannot be delivered over IP networks, then the potential benefits of this service do not matter. So we first look at the requirements in Section 4.2, and then discuss the issues and challenges in providing multimedia over IP in the light of these requirements (in Section 4.3). Next, we look at how some IP transport protocols might or might not meet the requirements (in Section 4.4).

4.2   Requirements

First, we consider the transport requirements for multimedia over IP streams and the coding requirements. Next, we consider what other network elements may be required for multimedia over IP. Finally, we examine signaling requirements.

4.2.1   QoS and Transport Requirements

As mentioned at the start of this chapter, two of the biggest challenges to providing sufficient QoS for multimedia over IP are related to delay: the end-to-end delay needs to be less than 400 ms, and the delay variance needs to be small as well. The delay variance is often called the jitter, referring to the small (hopefully!) fluctuations in arrival time of packets at the destination.

Unlike some other kinds of traffic (e.g., file transfer), voice and video traffic can survive occasional dropped packets. The encoding schemes are designed so that the occasional dropped packet results in a momentary slight degradation of quality and not catastrophic loss. For voice or video traffic, because they are real-time, retransmission of dropped or lost packets does not make sense, because by the time the retransmissions occurs, the arriving packet would be useless. (The stream would have moved on, so it would not make sense to play the few milliseconds of voice/video contained in the retransmitted packet.) This is quite different from the requirements for file transfer, for instance, where large end-to-end delay and arrival-time jitter can be tolerated as long as all packets eventually arrive at their destination.

Furthermore, for multimedia, a third requirement is that the streams can be synchronized at the receiver. Otherwise, the human at the receiver would be disconcerted by streams playing back in an out-of-sync manner. Therefore, the playback application must be able to synchronize packets from the different streams.

These requirements apply in both wired and wireless networks. However, they are generally harder to meet in wireless networks, where delay may be higher, and where higher error rates lead to higher rates of dropped packets, as we will discuss in Section 4.5.

4.2.2   Coding Requirements

As discussed earlier, even the PSTN has been switching over to digital encoding of voice, rather than using analog encoding, at least in the backbone of the network between switches. However, the PSTN is constrained to its 64-Kbps circuits, unless a massive redesign of the PSTN allows other rates. For multimedia over IP, there is no artificial 64-Kbps limitation. The question is whether significant bandwidth savings can be achieved for only modest loss of signal quality. The standard G.711, used in the PSTN, has a mean opinion score (MOS) of 4.3. The MOS is a scale from 1 (bad) to 5 (excellent) and is used to rate voice coders. We note that the G.729 coder, while scoring only 4.0 on the MOS test, uses only 8 Kbps of coding! The tradeoff is the added delay, 15 ms in the case of G.729 (it uses 10-ms frames and includes a 5-ms look-ahead buffer). In general, the most bandwidth-efficient coding schemes (like G.729) will need to do frame encoding. Frames typically range from 10 to 30 ms.

4.2.3 Other Network Elements

In many cases, only the two end hosts need to understand VoIP protocols and functions; the intermediate nodes in the path of the VoIP traffic need not add special functions for this support. They may have capabilities that can be used for VoIP support, such as QoS capabilities, but these are not specifically for VoIP support alone, unlike the functions in some of the new network elements we will now introduce. This is a good example of the end-to-end design principle introduced in Chapter 1. However, there are usage cases where it is helpful for intermediate nodes to perform special functions specifically to support VoIP applications. Examples of such functions and network elements include mixers, translators, and PSTN gateways.

4.2.3.1 Mixers

Imagine a conference call between two locations, in each of which two or more people speak to their colleagues at the other side through a speakerphone, and listen through the same speakerphone to what their colleagues say. Would you expect a separate connection to exist between each possible pairing of colleagues at the two locations? Of course not—the voices of everyone at one site are mixed and carried on one line to the other side, where the same thing happens in reverse.

Mixers help provide an analogous service in the case of VoIP telephony. Mixers combine two or more streams into one stream, where ideally the resultant stream is a superposition of the original streams. Mixing is a powerful concept with the flexibility to be used in a variety of scenarios. Unlike the speakerphone analogy, there need not be any notion of geographical proximity of the source streams. Like the speakerphone analogy, a mixer may be used to consume network resources more efficiently (than to transmit streams separately). This may not matter as much for high-bandwidth users, who can afford to receive several separate streams without a problem, but could be essential for low-bandwidth users, for example, users accessing the network through wireless links. If the low bandwidth is a severe problem, the encoding scheme of the output of the mixer could be chosen to optimize bandwidth usage. Mixers could also take the shape of other creative ideas, such as cleverly combining video scenes of different individual people to simulate a video scene of a group of people [1].

4.2.3.2 Translators

One of the usage scenarios of a mixer involves low-bandwidth users. Streams from other machines can be combined (mixed) into one. Now, imagine again the same low-bandwidth users, wanting to transmit to other users in the conference. If they had to transmit separate streams to each of the other users, they would again run into bandwidth problems. A mixer cannot help in this case, since the issue is not mixing multiple streams into one, but to transmit one stream and split it into multiple streams. However, a translator, placed on the other side of the low-bandwidth link, can help by taking one stream from such a low-bandwidth user and replicating it to transmit to multiple recipients. This translator usage is a kind of multicasting, in the sense of being an efficient way of transmitting from one to many. It is a kind of dual of the mixer going from many streams to one.

In general, though, a translator is not used solely for one-to-many translation. A translator can also take one input stream and produce one output stream, perhaps with a different encoding scheme. A useful rule of thumb to differentiate between mixers and translators is that given multiple input streams, the translator will process all of them separately, producing one or more output streams corresponding to each input stream, whereas the mixer would mix groups of two or more of the input streams to produce one output stream per group.

4.2.3.3   Combining Mixers and Translators

In general, there may be cascades of one or more mixers, and one or more translators, between the communicating end hosts. Figure 4.1 shows an example of mixers and translators in action. Notice that the symbols used for mixers and for translators indicate their functions. The various streams (represented by arrows) are mixed together (come together) in the center, for mixers, and not for translators. In mixer A, three streams are mixed into one, whereas in mixer B, two streams are mixed into one. In translator C, there are two input streams, each of which is translated into an output stream (perhaps to change the encoding scheme or other parameters). In translator D, we see translation from a multicast address to unicast streams.

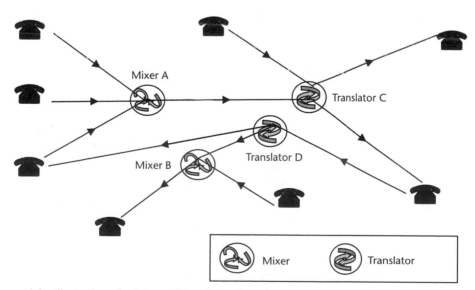

Figure 4.1   Illustration of mixers and translators in action.

4.2.3.4   PSTN Gateways

The traditional PSTN is going to be around for a long time, at least in some parts of the world. Many people will only be reachable through the PSTN, so being able to complete calls to the PSTN, or receive calls from it, is a very attractive system feature.[2] Various PSTN gateways have been designed for this purpose. The current leading standards-based approach is the media gateway approach, itself a consolidation of several proposals and currently standardized as H.248 (formerly H.GCP) by the ITU [4]. In this approach, two gateways are used for PSTN interworking: a *media gateway* for converting between signaling on both sides (e.g., between SIP and ISUP), as well as traffic encoding on both sides, and a *signaling gateway* for converting between signaling transport used on the IP side (e.g., SCTP) and signaling transport used on the PSTN side. The two gateways complement each other, one focusing on signaling and the other on signaling transport. The use of these two gateways, instead of just one integrated PSTN gateway, allows for more flexibility in deployment, and is a good example of modular design.

The media gateway itself is often split into a *media gateway controller* and a *media gateway*. The media gateway does the actual media conversion (e.g., between G.711 on the PSTN side and any of a number of possibilities on the IP side), while the media gateway controller controls the media gateway using H.248 signaling. Again, the split between media gateway controllers and media gateways is an example of modular system design. It allows each component to be upgraded separately, and each focuses on what it does and doing it well. Media gateways without control functions are cheaper, and multiple media gateways can be controlled by each media gateway controller. The arrangement of these network elements is shown in Figure 4.2. On the left side is the IP network, where the signaling traffic (e.g., SIP over UDP) and media streams (e.g., encoded with G.723) are split and go between the phone and media gateway controller, and between the phone and media

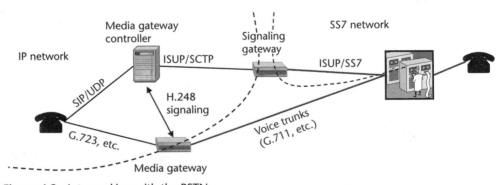

Figure 4.2   Interworking with the PSTN.

---

2.   Recall the principle from Chapter 1 that the incumbent technology (the PSTN in this case) needs to be respected, and backward compatibility or interworking is often useful, as it is here.

gateway, respectively. The signaling protocol conversion between SIP and ISUP signaling is done at the media gateway controller and the transport for the signaling changes from IP to the SS7 network, as it enters the SS7 network at the signaling gateway. Meanwhile, the media streams are translated (e.g., between G.723 and G.711), and changed from IP transport to placement on voice trunks in the PSTN, at the media gateway. H.248 signaling is used for control between the media gateway controller and the media gateway.

4.2.4   Signaling Requirements

Signaling between the two parties is required so that agreement can be reached on various questions, such as whether a session is desired in the first place and what kind of traffic it involves (e.g., video or audio and with what coders and decoders). There is also signaling involving other network elements, such as gateways to the PSTN when one or both parties are PSTN phones. In the early days of VoIP, the ITU's H.323 was the prominent signaling protocol. However, recently the IETF's SIP has grown in popularity and is replacing H.323 as the dominant signaling protocol for controlling VoIP sessions. One reason for this phenomenon is that SIP is "lightweight," specializing in session control signaling and doing it well, in the spirit of the piecewise specification approach mentioned in Chapter 1. Meanwhile H.323 is a broader specification and more cumbersome than SIP. We will discuss SIP and signaling requirements in detail in Chapter 5.

4.3   Issues and Challenges

Most of the challenges related to multimedia over IP have to do with QoS. Since the Internet provides only best-effort service, meeting the QoS requirements for multimedia traffic is a big challenge. Techniques for engineering an IP network for QoS differentiation can help, as will be discussed further in Chapter 7. With careful design, the end-to-end delay could be kept below 400 ms (or 300 ms or 250 ms, depending on the quality desired). Jitter could be handled by some buffering at the receiver, at the expense of slight increases in end-to-end delay.

High packetization overhead is another challenge. Because VoIP requires low latency, the transmitter cannot afford the luxury of waiting for many bits to accumulate before sending them together as a large packet. In fact, voice is typically sent in individual frames that are encoded frame by frame to achieve high compression efficiency. Each frame is typically 10 ms or 20 ms of speech, with the actual number of bits per frame depending on the bit rate. As explained in Section 4.2.2, the frame size and look-ahead buffer size add to the overall delay (10 + 5 = 15 ms for G.729, for example). Given the great savings in bandwidth usage, designers generally tolerate the delay for one frame. However, waiting for multiple frames to send together is a problem. Now, if we send frame by frame, the G.729 frame has only 80 bits, for example. The IP header is at least 20 bytes (160 bits) long, and it gets worse with IPv6 (40 bytes). Add to that the transport headers (UDP, RTP), and we see that the

Figure 4.3   Is there an ideal packet size for VoIP?

bulk of the packet is header overhead. Figure 4.3 illustrates the dilemma: On the one hand a heavy pressure is exerted to reduce the end-to-end delay by reducing the packet size. On the other hand, a heavy pressure push is exerted to increase the packet size to reduce the high overheads.

One approach to solving the packetization overhead problem is *RTP multiplexing*. RTP is a transport protocol optimized for real-time traffic [5, 6] (see Section 4.4.3). RTP multiplexing refers to sharing an RTP header between multiple sessions that are transmitted in the same packet.

Another approach to solving the packetization overhead problem is RTP header compression. RTP header compression can compress the IP/UDP/RTP header down to two to four bytes, from 40 bytes [7]. Its usefulness is limited by the fact that it works only for the single-link, unicast case. Otherwise, more header information is needed. Nevertheless, this is useful over a wireless link, where bandwidth is often scarce. In wireless links where packet loss is a problem, and where very high compression efficiency is required (because of limited bandwidth, for example), robust header compression may be used as an alternative to RTP header compression [8]. More recently, an enhancement to RFC header compression, Enhanced compressed RTP, has been developed, which is more robust for links where the delay is high and where packet loss and reordering may be encountered [9].

Additional issues and challenges related to multimedia over IP in wireless networks will be discussed in Section 4.5.

4.4   Transport Protocols

We briefly examine the suitability of traditional transport protocols like TCP and UDP for real-time multimedia applications, and then discuss at some length a transport protocol, RTP, designed with such applications in mind.

4.4.1   TCP

TCP is the main transport protocol used in IP networks. However, TCP was designed for traditional data network applications, like file transfer, with different transport requirements from multimedia applications, as explained in Section

4.2.1. For example, TCP uses ACKs so that it can retransmit packets not received by the receiver. However, for real-time multimedia traffic, this is not useful, since the application would have moved on to play the most recently received packets by the time a retransmitted packet arrives, and so would not have use for retransmitted packets.

4.4.2   UDP

Whereas TCP is a complex protocol with several mechanisms to support traditional applications like file transfer, UDP is a more lightweight protocol that is more suitable for real-time multimedia applications. For example, UDP neither detects unreceived packets, nor arranges for them to be retransmitted. However, while not retransmitting unreceived packets is good for real-time multimedia applications, it would be helpful to detect them, as well as provide some time-stamping capabilities. So UDP could be used, if it could be augmented in some ways.

4.4.3   RTP

RTP was designed from the ground up to carry real-time traffic. Therefore, it has features that help in providing the QoS needed for multimedia over IP. These features include time stamping, the ability to synchronize different streams, and the ability to support a mixer. The companion protocol, Real-time Control Protocol (RTCP), is used with RTP to distribute reports on the performance of the real-time data transport to the participants in the conference.

   RTP is typically run over UDP (i.e., RTP packets, including the RTP headers, are put into UDP packets as the payloads), where RTP uses even port numbers, and the corresponding RTCP packets use odd port numbers that are one higher than the port numbers used by RTP. We will first explain how RTP works and then describe RTCP. RTP over UDP can be thought of as augmenting the UDP header information with additional information useful for real-time traffic, and placing these additional bits of information in the front portion of the UDP payload. What are these bits of information, and how do they help? We examine the RTP header, shown in Figure 4.4 and highlight the fields most pertinent to facilitating real time traffic. In particular, we will discuss:

- Time stamp;
- Sequence number;
- Synchronization source (SSRC) identifier;
- CSRC identifiers and CSRC count.

We first note that two notions of packet sequencing are needed. In order to facilitate synchronization of different media, and to facilitate jitter computations (for jitter monitoring by RTCP and possible correction by the application), highly accurate time stamping is needed. This is analogous to a postal worker stamping the date on the stamp on an envelope, where the recipient of the letter will know when

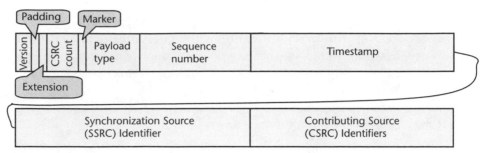

Figure 4.4 RTP header.

the letter was sent. In the case of RTP time stamping, the time resolution is of course finer than in the postal analogy, and the value is contained in the *timestamp* field (an example of the use of the time stamp will be shown shortly). The second notion of packet sequencing arises from the observation that it would not necessarily be trivial for the receiver to figure out if any packets have been lost, by just looking at the time stamps of the received packets. Hence, a *sequence number* is also used, where the sequence number is incremented by one each time an RTP packet is sent. If there is a break in the consecutive sequence numbers received, a receiver can detect that packets have been lost. The sequence number also allows a receiver to correctly reorder packets that have arrived out of sequence.

What if there are multiple streams from different sources? How would the receiver tell them apart? The SSRC field is used to allow differentiation of packets associated with different synchronization sources, where a synchronization source may be, for example, a microphone or camera. Sequence numbers and time stamps have relevance for packets from the same synchronization source. In other words, packets from the same synchronization source need to be played back in sync, and the SSRC together with the time stamps and sequence numbers allows an application to do so. There may be multiple streams, each with their own SSRC identifier, in one RTP session, an example being multiple video cameras, each producing a stream. Furthermore, there can be one or more RTP sessions in a general multimedia session.

The CSRC list is useful only when the stream has gone through a mixer. The CSRC list is the list of contributing sources, in other words, the SSRCs of the streams that went into the mixer. The SSRC of the packets coming out of the mixer is the SSRC of the mixer. Hence, the CSRC list allows receivers to identify the contributing sources despite the streams having been mixed. An application could then display the list of speakers.

Why does not RTP do more to guarantee timely delivery? Since it is a transport protocol and not a lower-layer protocol, the uses of RTP are limited, as with any other transport protocol. RTP does provide tools (such as time stamps) for an application to play media streams properly synchronized with segments in the right sequence and time order. Further tools for applications to monitor overall performance are provided by the companion protocol, RTCP. Table 4.1 summarizes what RTP does and does not do.

Table 4.1   Summary of Capabilities of RTP

What RTP Does Do	*What RTP Does Not Do*
End-to-end transport	Guarantee timely delivery
Provides tools like time stamping, sequence numbers, SSRC that multimedia applications can use	Deal with resource reservation

4.4.3.1   RTCP

RTCP distributes reports on the performance of the real-time data transport provided by RTP. The reports from receivers can be used by senders to dynamically adjust their encoders, or for flow and congestion control. Also, a network service provider might set up a node to receive RTCP packets to monitor the network for network problems. Such a node, or any other node that receives the various RTCP feedback reports, could assess the scale of problems. For example, if the quality of real-time data transport received is very bad only at a couple of nodes clustered together, but not at other nodes, then the problem may be suspected to be local. However, if all nodes report roughly the same performance problems, a global problem is more likely.

RTCP provides for both sender reports (SRs) and receiver reports (RRs). SRs contain statistics for *both* transmission and reception. In that case, why are RRs necessary? RRs are used by nodes that are not active senders. The concept of active sender in this context applies only to the time interval since a node issued its last report. If a node has sent RTP packets since then, it issues an SR. Otherwise the node issues an RR. The other use of RRs is in the case that an active sender is reporting on packets received from more than 31 sources. The first 31 packets will be reported in the SR and the rest in one or more additional RRs sent with the SR. The main difference between the SR and RR packets is that the SR includes a 20-byte sender information section. This sender information section details the packet count transmitted by the sender (a running count for as long as the same SSRC identifier is used). An NTP time stamp is included too, so that when other receivers send their reports, round-trip propagation can be estimated.

What information do the RRs contain, and how might they be useful? The reports are organized by SSRCs. For each SSRC, fraction lost, cumulative number of packets lost, highest sequence number received, interarrival jitter, the last SR time stamp received, and the time since that last SR, are sent. The fraction lost is an estimate of the proportion of packets lost to received packets, during the most recent reporting period, whereas the cumulative number of packets lost is an absolute number. A jitter estimate can be computed, as specified in the RTP specification [5]. The SR time-stamp information will assist the sender in computing round-trip propagation time. Note that the fraction lost and the jitter estimate provide estimates of network congestion that could be useful to the application.

4.5   Wireless Multimedia over IP

If one or more of the links in the communication path are wireless links, additional challenges arise. Wireless links have lower bandwidths than wired links, making schemes for compression of RTP headers more necessary in the wireless case. Wireless links generally have higher error rates than wired links. The "raw" error rates (without error control coding) can be quite high, and even after the application of forward error correction (FEC) codes, the error rates are still higher than in wired links. The high error rates have implications on multimedia over IP in several ways. First, header compression is very important for multimedia over IP applications, as explained earlier, because of the short packets and large header overheads of uncompressed headers. However, compressed headers are very sensitive to errors, so special header compression algorithms should be used. Second, certain voice codecs perform very poorly, and are therefore less suitable for use, in high-error environments like wireless links. Special, more robust algorithms like 3G-324M should be used over wireless links.

While we will discuss QoS in greater detail in Chapter 7, we note here that wireless links can add significant amounts of additional delay to the end-to-end delay. For delay-sensitive applications like conversational voice and video, there is already a challenge providing satisfactory delay characteristics when the entire network path is wireline. With wireless links in the path, the challenge is increased. Another phenomenon in mobile wireless networks is handoffs, as mobile nodes move from one connection point to another. Handoffs typically result in additional latency, even if care is taken to avoid packet loss through buffering schemes. Schemes that buffer packets and retransmit them after the handoff is complete may work well for regular data traffic, but the latency is a more serious problem for multimedia over IP. We will see some schemes for reducing handoff latency later in this book.

Another issue is battery power consumption. Wireless devices, being small and mobile, rely on batteries for power, so it is desirable to minimize power consumption, and thus to maximize the length of time the device can operate before needing to be recharged. Traditional cellular phones come with sophisticated power-saving mechanisms. For example, as we saw in Chapter 3, location update and paging concepts work together to effect power savings. GSM also has a variety of other mechanisms for power conservation. GPRS similarly has power-saving mechanisms, including the use of three mobility management states (discussed in Chapter 11). However, 802.11 WLAN does not have the same range of power-saving mechanisms, because it is a link-layer protocol. Thus, it cannot take advantages of mechanisms, such as the state-dependent location update frequency in GSM, that make use of cross-layer interactions. The standard 802.11 has no understanding of the traffic it carries, whereas GSM can associate the application-level concept of being in a call or not being in a call to the frequency of location updates as performed in the network layer. Therefore, in the case of voice over WLAN, vendors must implement their own non-standard power conservation schemes. In fact, vendors are doing so, although one could argue that a standardized solution might have been preferable [10].

4.6 Summary

This chapter and the next focus on technologies that directly support conversational voice and video applications, which are expected to play an important role in the future wireless Internet. This chapter deals with transport aspects and RTP in particular. After discussing the motivations for multimedia over IP, we surveyed the requirements (e.g., QoS and transport requirements and coding requirements) and discussed various network elements helpful for delivery of multimedia over IP traffic, such as mixers, translators, and PSTN gateways. We discussed issues and challenges for transporting multimedia over IP and over wireless links in particular. The basics of RTP and RTCP were explained.

References

[1] 3GPP TS 23.107, "Quality of Service (QoS) Concept and Architecture (Release 5)," V5.9.0, June 2003.

[2] Hassan, M., A. Nayandoro, and M. Atiquzzaman, "Internet Telephony: Services, Technical Challenges, and Products," *IEEE Communications Magazine,* April 2000.

[3] Schulzrinne, H., A. Rao, and R. Lanphier, "Real Time Streaming Protocol (RTSP)," RFC 2326, April 1998.

[4] ITU Recommendation H.248, "Gateway Control Protocol," 2002.

[5] Schulzrinne, H., et al., "RTP: A Transport Protocol for Real-Time Applications," IETF RFC 3550, July 2003.

[6] Schulzrinne, H., and S. Casner, "RTP Profiles for Audio and Video Conferences with Minimal Control," IETF RFC 3551, July 2003.

[7] Casner, S., and V. Jacobson, "Compressing IP/UDP/RTP Headers for Low-Speed Serial Links," IETF RFC 2508, February 1999.

[8] Bormann, C., et al., "Robust Header Compression (ROHC). Framework and Four Profiles: RTP, UDP, ESP, and Uncompressed," IETF RFC 3095, July 2001.

[9] Koren, T., et al., "Enhanced Compressed RTP (CRTP) for Links with High Delay, Packet Loss and Reordering," IETF RFC 3545, July 2003.

[10] Wexler, J., "Preserving Wireless VoIP Handset Battery Life," http://www.nwfusion.com/newsletters/wireless/2004/0419wireless1.html, 2004.

CHAPTER 5

Session Initiation Protocol (SIP)

For many Internet applications, two or more hosts form temporary associations for the exchange of data. For example, FTP associates two hosts for the purpose of file transfer, in a file transfer session. In a multicast video streaming session, the video transmission source streams video to multiple recipients using multicasting. In order to establish the associations and the parameters for data exchange, session initiation and management mechanisms are needed. SIP is one such session initiation and management protocol that is general enough to not depend on the type of session it initiates and manages, or on the underlying transport protocol [1]. It is a text-based protocol that borrows much of its syntax from HTTP and consists of requests and responses like HTTP.

SIP, despite its name, is used not just for initiation of sessions but also for management and termination of sessions. SIP is used for managing a session in the sense that it can add or drop participants to and from a session, or change parameters such as media encodings during a session. SIP is especially good for initiation, management, and termination of real-time multimedia sessions. For some other kinds of session management protocols, such as file transfer sessions using FTP, only a particular kind of session is supported. The user can specify some parameters related to file transfers (binary or ASCII file transfer), and other IP-based systems like DNS are consulted as needed to locate machines with which a file transfer session is desired. However, SIP offers much more flexibility. In addition to supporting features for multiparty sessions, it supports powerful and flexible ways to describe and negotiate session parameters, and also includes powerful location mechanisms to find machines to invite to sessions.

Despite the fact that SIP is designed as a general-purpose protocol, in this chapter we focus on its use for real-time multimedia sessions because this is the area in which most SIP development has occurred, and because this book is about wireless Internet telecommunications. But first, why bother with a new signaling protocol when the telephony signaling protocols already exist? Why not just reuse the telephony signaling protocols and adapt them as necessary for working over an IP network?

An IP network is different from a telephone network. Not everything that pertains to the setting up and handling of circuits in a telephone network has an analog in VoIP, nor is this necessarily desirable. An example might be telephone network signaling references to a particular circuit in a trunk (a collection of circuits). Also, in a telephone network, the switches handle the call setup and maintenance signaling between the end users, as well as the voice circuits that are used are between these same switches. In other words, there is no difference between the signaling path[1] and the path that the end-to-end circuit takes. On the other hand, in an IP network, the path of packets between the two end users is often directly between the two, bypassing the signaling elements in between (see Figure 5.1 for an illustration of how packet switching has this flexibility). This is because an IP network has at its heart the IP packet-switching protocols that cause packets to be eventually delivered to their destination addresses, whereas circuits need to be set up in the telephone network, and the end points of these circuits are switches and phones. Furthermore, the IP network will only deliver packets between two end points, without guaranteeing that the packets will take any particular path.

IP networks have more to offer than just the transport of VoIP packets. Popular services like Web browsing and e-mail are also supported by IP protocols. One powerful reason for creating a new signaling protocol for IP telephony is that it facilitates the integration of telephony with other IP-based services, sometimes known as *CTI*. For example, some links on a shop's Web page could cause a multimedia phone call to be made to the customer service department, bringing the customer in more immediate, real-time contact with a person who could answer specific questions. SIP

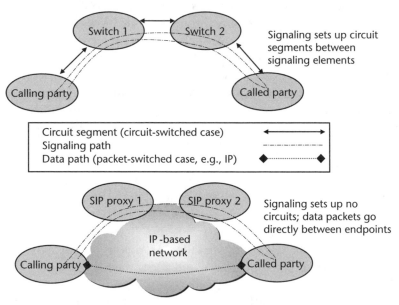

Figure 5.1 Signaling and data traffic need not take the same path in an IP network, unlike in a circuit-switched telephone network.

---

1.   Strictly speaking, the signaling between any two switches in the path between two phones actually goes on a separate SS7 network, whereas the circuit that is set up is carried in a trunk between the two switches. However, the point is that the pair of switches that are the end points for each and every circuit segment is the same pair that was involved in signaling for setting up that circuit segment in the first place.

was designed as an Internet-style protocol. Although certain session establishment functions are similar to functions used in the traditional PSTN signaling schemes, the whole packaging and design makes it very different. From being a text-based HTTP-like query-response protocol to the modular nature of the protocol design (it doesn't try to do everything, but reuses other IP protocols for QoS and real-time transport), it has the flavor of an IP-based protocol.

5.1   Requirements for Session Initiation

5.1.1   Basic Requirements and Characteristics

We begin with an analogy: Suppose there are two countries, country A and country B (in the following, referred to as A and B for short), which are separated by an ocean. One day, A decides it would like to trade goods with B, say fruits from A for vegetables from B. A's trade minister sends a messenger with a message to B's trade minister, to invite B into the proposed new trading relationship with A. There may be negotiation of terms and conditions. Supposing that B's trade minister declines, then the trading relationship is not established. Supposing instead that B's trade minister accepts, then the trading relationship is agreed upon. Both sides then set in motion a process whereby fruits from A are collected and shipped from the port in A over the ocean to the port in B, and vegetables from B are likewise collected and shipped from the port in B to the one in A.

This describes, at a basic level, the process of initiation of a trading session between A and B. The initiation of a telecommunication session is similar. Whether it is the initiation of a traditional phone conversation session, or a multimedia over IP session, the basic requirements and characteristics are the same as for initiating a trading session. Table 5.1 shows the basic requirements and characteristics of session initiation, and how it is manifested in different examples of session initiation.

Table 5.1   Basic Requirements and Characteristics

Requirements and Characteristics	Trading Session	Traditional Telephone	SIP for Voice over IP
One party initiates, inviting the other party	Trade minister of A sends a message through a messenger to counterpart in B	Phone A signals to phone B the invitation through the phone network	Phone A signals with a message to phone B over the IP network
There may be negotiation of terms, conditions, and/or parameters	The messenger from A negotiates with his/her counterparts in B	Typically the decision is either accept or decline with no negotiations	Negotiations of parameters takes place, using SDP information from invitation
The second party may decline	The messenger is ignored or negotiations fail	Phone B is not picked up	Phone B is not picked up, or negotiations fail
The second party may accept	An acceptance message is returned through the messenger	Phone B is picked up; no further signal is sent to phone A	Phone B is picked up; an acknowledgment message is sent to phone A

5.1.2 Additional Requirements

The analogy can be extended in several ways. How does A's trade minister locate B to initiate the session in the first place? This is the problem of locating the other party to initiate a session. Furthermore, whereas country B would presumably never change its location, SIP offers convenience to its users, and more challenges to itself, by supporting location of users even if they move from time to time. Now, suppose that ocean travel is not free, and that ocean shipping lanes are regulated in some way. Can the traffic be made to pass through certain checkpoints (perhaps at some islands in the ocean) along the way for accounting, billing, and other administrative purposes? The analogous requirements are supported by SIP.

5.2 Fundamentals of SIP

Figure 5.2 shows a simple case of the use of SIP to initiate and terminate a session between two IP phones (by IP phone, we mean a "phone" that can send and receive multimedia traffic over IP[2] and that uses an IP-friendly protocol like SIP for managing the sessions). The initiating side invites the other side, with an INVITE request message. Before the person accepts the call (e.g., by picking up the phone), a "180 Ringing" message is sent to indicate that "ringing" (or alerting) is ongoing. This allows the calling side phone to provide *ringback* (ringing tones on the calling side to indicate that the other party is currently being alerted). Note that the 180 Ringing message is our first example of a *provisional response* in SIP. Provisional responses are typically informative messages that do not require an acknowledgment because

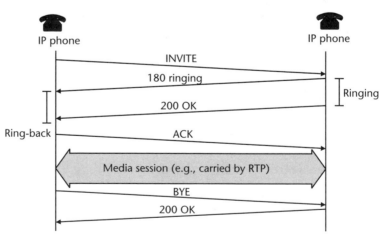

Figure 5.2   A basic SIP-managed session.

---

2.   Although perhaps limited, for instance, to voice traffic only.

it is not essential that the other side receives them. The "180" is the numeric *response code*.

When the called side accepts the call, a "200 OK" message is sent back. Unlike the provisional responses like 180 Ringing, some responses are *final responses* that require acknowledgment. The 200 OK message is a final response. The "200" is the numeric *response code*. We will explain the difference between provisional and final responses, and the various response codes in Section 5.3.2. For now, we note that the 200 OK message responding to the INVITE means all is okay, and the session initiation is complete. After the calling party receives the 200 OK, it acknowledges receipt with an ACK message. Note that this exchange of three messages (INVITE, 200 OK, and ACK) is sometimes referred as the *three-way* handshake of SIP. All the necessary session initiation information has been exchanged, so the two parties can begin a multimedia session.

Finally, one party hangs up, and sends a BYE message to the other party. In the example depicted in Figure 5.2, it so happens that the initiator of the session is also the one that wants to end it. It could also be that it is the other party that sends the BYE message. In either case, a 200 OK response is returned by the recipient of the BYE message to indicate successful compliance with the request.

In going through the preceding example, some or all of the following questions may come to mind, relating to how SIP handles requirements on functionality:

- How does the initiating party find the other party? Does it have to know the IP address of the other party? The answer can be found in Sections 5.2.1 and 5.2.2.
- In a telephone call, both phones should have the same basic capabilities to handle analog voice encoded on copper wire in the band 3 kHz and below. There is no need to negotiate parameters. With voice and video over IP, each of which uses a variety of different coders and decoders, the situation is different. The two parties must have a way to agree on the media and the coding schemes. Can SIP handle this? How? The answer can be found in Section 5.2.3.
- In the example shown in Figure 5.2, the SIP signaling is totally end-to-end; in other words, it goes directly between the two parties. Is this viable in a commercial or even an enterprise setting? In a commercial setting, how can the "network" keep track of customer usage to bill them accordingly? The answer can be found in Section 5.2.1, in the proxies that will be introduced, and the optional Record-Route field in the SIP header.

In order to provide a framework into which to put the occasional mention of specific SIP headers (as we will do so in discussing various features), we illustrate an example next. This is how the INVITE message header (of the INVITE message in Figure 5.2) might look. As mentioned earlier, the header is text-based. Also, there is a message body (not shown here) that uses SDP to specify session parameters.

```
INVITE sip:gandalf@lotr.com SIP/2.0
Via: SIP/2.0/UDP hobbit.shire.org;branch=zlkjfdslkg89Ug3
Max-Forwards: 70
From: Frodo <sip:frodo@shire.org>;tag=5478921342
To: Gandalf <sip:gandalf@middleearth.org>
Call-ID: hl3f432fklj3@hobbit.shire.org
CSeq: 1 INVITE
Contact: <sip:frodo@hobbit.shire.org>
Subject: the ring
Content-Type: application/sdp
```

Notice the first line contains the *method* name, in this case, INVITE. The other lines contain header fields and their values. The meanings and uses of these and other fields are briefly summarized in Table 5.2. Further details will be provided in the discussions on SIP functionality in the rest of this chapter. A complete list of header fields can be found in the SIP specification [1]. More detailed examples of basic SIP call flows, including the SIP header and message details for each message, can be found in Johnston et al. [2].

5.2.1   Locating Other SIP Users

In the traditional telephony world, when we call someone, we need to know that person's telephone number (not to mention prefixes like country codes and area codes, or to obtain International Direct Dial services). For IP telephony, do we similarly need to know the IP address of the machine used by the person we wish to call? No, we don't! SIP provides a naming scheme that is more human-friendly than IP addresses. The format is: *user@host*.

Table 5.2   Selected SIP Header Fields

Call-ID	A unique ID for a call, related to a particular invitation.
Contact	This field has several uses. In the context of the INVITE message, it tells the other party how to directly reach the initiating party in the future.
Content-Type	Describes the content type of the message body, which is typically application/sdp, indicating that it contains session description parameters in the SDP format (to be discussed later).
CSeq	A sequence number that allows ordering of transactions in a dialog.
From	Contains the initiator of the request.
Max-Forwards	Like a TTL for SIP forwarding between proxies (proxies will be introduced later in the chapter).
Record-Route	The use of this field will be introduced later in this chapter.
Requires	An optional field that indicates SIP extensions that the user agents (UAs; see Section 5.3) must understand to process the request.
Route	Used to force routing of the request through a set of proxies (to be discussed later).
Subject	Subject of the call.
To	Specifies the recipient of the call.
Via	Shows the path taken so far (if a request goes through multiple proxies, each will add themselves to the header, by adding a Via header; so there will be multiple Via headers). This allows all the signaling for a transaction to follow the same path. Will be discussed later.

As can be seen, the SIP naming scheme bears family resemblance to other IP-style names, such as e-mail addresses (user@host). The "host" portion of the name is often a domain. Whether it is a domain name or machine name, it is recommended that the name be specified as a fully qualified domain name (FQDN), although IP addresses are also acceptable. We note that as for e-mail addresses, SIP includes user names and not just host names. This allows the desired party to be specified more precisely than just to the level of the machine. In traditional telephony, in comparison, only the phone number is available. There is no concept of calling a particular user at a phone number, but just a phone number, which really refers to a phone line (of course, we know how to verbally ask for a particular person when we dial a multiperson residence or office).

It may appear that using a human-friendly naming scheme, while more convenient than numeric IP addresses, does not actually solve the problem of locating the desired party, but only adds an extra database lookup (to find the IP address corresponding to the host). However, there is a solution. The solution has two main ideas:

- Two types of SIP network elements assist in delivering an INVITE message from the sender to the desired recipient. These SIP network elements are the SIP proxies[3] and the redirect servers. A SIP proxy receives an INVITE message (and possibly other SIP messages as well) and forwards it "nearer" to the destination, perhaps directly to the destination itself, if it knows the corresponding IP address. Since it is doing this on behalf of the initiating party, it is aptly called a SIP proxy. Meanwhile, a redirect server will not forward SIP messages as a SIP proxy would. However, it does something just as valuable. It returns a redirection to another server or proxy that may be more knowledgeable about finding the user. SIP proxies and redirect servers can be chained (an INVITE message could go through multiples of each type before arriving at its destination) in any order and any number of times. The question arises: How do SIP proxies and redirect servers know where to proxy forward or redirect a given message? The second idea contains the key.

- SIP relies on the notion of a *location service* that provides SIP address bindings for a particular domain. A SIP address binding is a mapping between an *address-of-record* URI and one or more *contact addresses*. By providing the bindings for a particular domain, we mean that the location service provides SIP address bindings for anybody in that domain. Table 5.3 shows an example of bindings available from a location service for the domain must.edu.my. In this example, we suppose that itd.must.edu.my is a subdomain of must.edu.my, such as a large department within a university, and with its own location service for itd.must.edu.my. So we could imagine a

---

3.   SIP proxies are also sometimes called proxy servers. However, we prefer to use the term "SIP proxy," to avoid confusion with the SIP UAS concept. On the other hand, a redirect server is properly called a redirect server in the sense that it is a special case of a UAS.

Table 5.3 Location Service Mappings Example

"Map From" URI	*"Map To" URI*
dwong@must.edu.my	dwong@itd.must.edu.my
temp@must.edu.my	temp@mit.edu
somebody@must.edu.my	somebody@192.168.1.41

SIP proxy for the must.edu.my domain receiving an INVITE for sip:dwong@must.edu.my, obtaining the binding from the must.edu.my location service, and forwarding the INVITE to the SIP proxy for itd.must.edu.my as the next link in the chain. The second entry shows a case where the binding for a user is actually outside the domain. SIP is flexible enough to handle this kind of binding, allowing forwarding and mobility (to be discussed in more detail later). The third entry is an instance of when the location service has the actual IP address of a machine where the user is reachable.

We expand the call flow of Figure 5.2 in Figure 5.3, with the addition of two SIP proxies. The calling SIP phone in this case does not know the current IP address of the SIP phone it is calling, since it may know only the name of the other party. Hence, the calling SIP phone requires help from one or more SIP proxies and redirect servers. The figure illustrates a case in which the signaling goes through two proxy servers. For example, the first proxy may be the one handling the domain of the SIP user being called (e.g., must.edu.my), and the second may be a proxy at another location where the user is currently located (e.g., mit.edu).

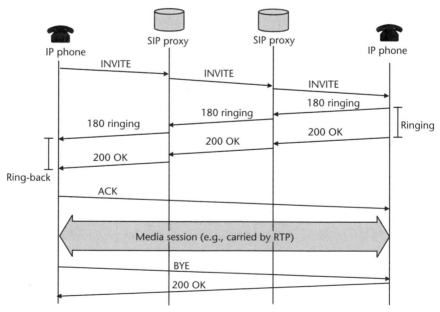

Figure 5.3 Using SIP proxies.

It can be observed that the INVITE message and responses to it all go through both proxies, but that subsequent messages (from the ACK onwards) bypass the proxies (shown on the figure by direct straight lines between the two end hosts). This is because, after the initial INVITE transaction, both SIP phones know the IP address of the other side, so they do not need to keep signaling through the proxies. However, sometimes it is desirable that, for administrative purposes, some or all of the proxies remain in the signaling path throughout the lifetime of the session. SIP supports this feature as well, using a *Record-Route* field in the SIP header. Proxies that want to remain in the signaling path during the session can add themselves to the *Record-Route* field in the INVITE message header, and their request will be honored. The *Route* field can be used in subsequent transactions to force routing through those proxies. In commercial settings, this makes proxies a natural place to collect billing records, and for other aspects of the sessions that the service provider wishes to control. In Figure 5.3, we have not yet shown the location service. However, it can be seen later. SIP UAs can also be configured so they will always initiate sessions through a particular proxy, regardless of whether they know how to contact the other party directly. Such a proxy is called an *outbound proxy,* and can be useful for control and administrative purposes.

The next example, shown in Figure 5.4, employs a redirect server. In this example, the redirection is to a SIP proxy that forwards the INVITE to the desired destination. However, redirection could also be directly to the desired destination. Unlike a proxy, a redirect server does not stay in the path of the signaling (not even for the rest of the INVITE transaction). Furthermore, it is up to the recipient of the

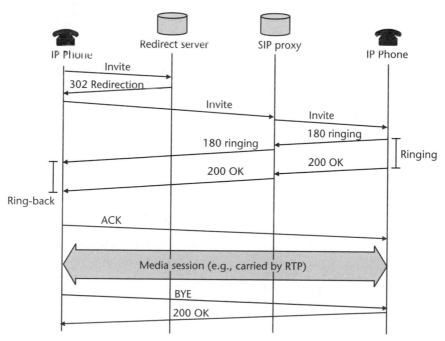

Figure 5.4 Session initiation with redirection.

redirect response to resend its request to the location provided by the redirection. Since there are several types of redirects, they are distinguished by response codes. The 302 Redirect is a "temporary" redirection, as compared to the 301 Redirection that is "permanent." Information provided in a permanent redirection should be cached and used to update local address books. Note that proxies and redirect servers are just logical roles, which means although they need not be implemented in separate machines in the network, they are still logically distinct roles even if implemented on the same machine. A person's SIP phone may act as a proxy or redirect server when that person wants to enable call forwarding to his or her current location.

How does a location service know the correct address bindings for all SIP users in its domain? Information may be entered into a database by a system administrator, or some other provisioning process based on external knowledge may be used. Another method is to use the SIP registration procedure with the REGISTER message, as will be explained in Section 5.2.2.

Another benefit of separating the concepts of user and machines is to facilitate the feature that allows a person to "move" from machine to machine. This idea of *personal mobility* will be explored further in Chapter 6, as well as more general mobility management solutions that allow machines that move to be contacted at their new locations for services other than receiving SIP messages.

5.2.2 SIP Registration

SIP registration allows an SIP UA to keep a location service current regarding its location. SIP registration requires a SIP *registrar*. Like the redirect server, the registrar is a special user agent server (UAS) designed for a particular purpose—the registrar acts

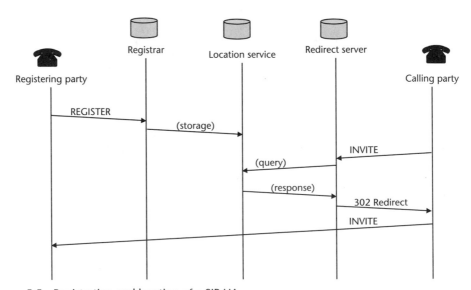

Figure 5.5 Registration and location of a SIP UA.

as a front end to the location service for updating the address bindings. The SIP proxy can be thought of as a dual of the registrar, as it obtains address bindings from the location service. Since the registrar and the proxy are merely logical roles, they could be implemented in one element, which could also have a database with the location service.

We illustrate the SIP registration procedure with an example, in Figure 5.5. In the example, we show how a SIP UA registers its presence in a particular location, and how this allows another SIP UA to find it to invite it to a session. In the call flow, we show only up to the INVITE reaching the SIP UA, the rest of the call flow being the regular flow. Also, by the proximity in time of the INVITE arriving at the proxy soon after the registration data is stored with the location service, we do not mean to imply that this data must necessarily be queried soon after the registration data is stored. In fact, registration data is valid for as long as the value specified in the "Expires" header in the REGISTER message. This may be as long as 136 years.

In the figure, the three logical roles of registrar, location service, and SIP proxy are drawn separately for illustrative purposes, although they could also be implemented in one machine. The SIP specifications mandate very little about the implementation of the location service: namely that there must be a registrar for a domain that can read and write to and from the location service, and a proxy in the same domain that can read from it. Therefore, we used generic messages to show its interactions with the registrar and proxy.

5.2.3 Session Parameter Negotiation

Most of the information related to session initiation, management, and termination is contained in the SIP header. The notable exception is the session description (the types of media and the encodings), which is carried in the body rather than in the header. Typically, Session Description Protocol (SDP) is used. SDP was originally designed for announcements of multicast multimedia streaming sessions, but is general enough to be usable with SIP with minor modifications. In Section 5.4, we explain SDP further, as well as how SIP uses it to offer session negotiation capabilities.

5.2.4 SIP for Telephony and PSTN Interworking

Obviously, SIP is not going to "take over the world" of voice telephony right away. The PSTN will be around for a long time. The value of SIP will be greatly enhanced if it can interwork with the PSTN so that SIP phones can make calls to PSTN phones and vice versa. As we discussed in Chapter 1, it is very important to respect the incumbent (the PSTN in this case), for otherwise the transition to a new technology is difficult, even if the new technology is otherwise very attractive. Work on such practical issues has been done in an IETF working group that studies applications of SIP (the "sipping" working group, with sipping standing for Session Initiation Proposal Investigation) [3, 4].

5.3   Digging Deeper

SIP employs a client-server architecture, where the roles of a client or of a server are neither permanent nor necessarily played by different network elements. In fact, the IP phones at the two ends in Figure 5.2 would each most likely be able to play both client and server roles as needed. They would contain software with SIP UA functionality. A SIP UA has both *user agent server* (UAS) and *user agent client* (UAC) functionality, so it can act as a UAS or as a UAC as necessary. For example, in the first *transaction*, where one phone invited the other to the session, the UA in the inviting party acts as UAC and the UA in the recipient of the INVITE message acts as a UAS.

Other than the client and server roles that UAs play, SIP also involves proxies and redirect servers. We note that these are also logical roles and any given network element is not constrained to assuming just one of these roles. Similarly, the SIP registrar is also a logical role.

SIP uses only two types of messages, namely, *requests* and *responses*. Requests are messages from clients to servers, and responses are messages from server to client. In fact, since SIP was designed to have a very HTTP-like syntax, the SIP requests and responses borrow much of the syntax of HTTP requests and responses. For example, the request messages for both protocols start with "Request-Line" and the response messages with "Status-Line." Response codes are put on the status line of both protocols. HTTP has five classes of response codes; SIP has practically the same five classes (with the same numbering), plus an additional class (the 600 series). Requests and responses will be discussed in more detail in Sections 5.3.1 and 5.3.2, respectively.

The reader may have noticed that we have talked about SIP transactions, such as the INVITE transaction, without defining a transaction. A SIP transaction is an exchange of messages between a client and a server beginning from the first request that the client sends and ending with the final response (in contrast to a provisional response; see Section 5.3.2) that the server sends. In the special case of an INVITE transaction, the transaction may or may not include the entire three-way handshake. If the final response to the INVITE is a 200-series response (see Section 5.3.2), the ACK is considered a separate transaction. Otherwise, the ACK is included in the same transaction.

How do SIP UAs maintain the concept of a session? For example, after a session is established, if a BYE message is subsequently sent, how does the recipient of the BYE message know which session the BYE is intended for? We need another concept, the SIP *dialog*. A dialog is a temporary peer-to-peer relationship between two SIP UAs. Its former name (used in earlier stages of the development of SIP) is *call leg*. A dialog is identified by a dialog ID, where the dialog ID comprises the Call ID, a local tag, and a remote tag. The local tag is the tag in the From header field, and the remote tag is the tag in the To header field. The local and remote tags are reversed at the other side.

5.3.1 Requests

SIP requests are sent from a SIP client to a SIP server and are meant to invoke an operation on the server. Six requests are defined in the base SIP specification, namely, the six *methods* REGISTER, INVITE, ACK, CANCEL, BYE, and OPTIONS. The method relates to the operation that a request message is meant to invoke on the server. The base methods are the following:

- REGISTER: as part of registration procedure;
- INVITE: as part of session initiation procedure;
- ACK: as part of session initiation procedure;
- CANCEL: as part of session initiation procedure;
- BYE: as part of session termination procedure;
- OPTIONS: as part of capability querying.

We have seen four of these methods so far, with CANCEL and OPTIONS remaining.

Traditional telephony uses the notion of hanging up a call. We have seen how BYE can be sent analogously to terminate a SIP session. However, the notion of hanging up extends also to the time before the circuits are all established. In the SIP case, this would be the time before the INVITE transaction is completed. How can we indicate a desire not to proceed after sending the INVITE but before receiving a final response to it? The CANCEL method, which can be used to cancel an INVITE transaction, is the solution. In fact, the CANCEL method can be used by a UAC to cancel *any* request previously issued by the UAC, by requesting that the destination UAS stop processing the previous request and respond to it with an error response. However, CANCEL is only effective when a final response has not yet been issued to the previous request. In practice, therefore, CANCEL is most useful to cancel INVITE transactions, since the UAS may take a long time to issue a final response (the UAS must wait for the other side to "pick up the phone"). Figure 5.6 shows an example of the use of the CANCEL message.

Sometimes, having a query mechanism is helpful, to discover capabilities of a UA or proxy without needing to invite that machine to a session. The PSTN world has no good analog for this, as the PSTN does not have a range of coders and other options to choose from. However, in IP telephony, there is a range of coders, supported methods (beyond the basic six), and extensions. The OPTIONS method allows for such querying. For example, a UA may want to query another UA or a proxy on whether it supports certain SIP extensions, rather than waiting to send an INVITE that requires a certain SIP extension that may not be supported.

5.3.2 Responses

SIP response messages are responses from UA servers to UA clients. They provide information to the requesting UACs on how the response has been, or is being,

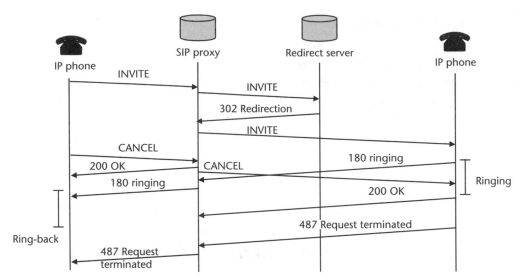

Figure 5.6   Use of the CANCEL method to cancel an INVITE.

handled. All SIP responses provide a status code and a reason phrase. The status code is a three-digit integer that has a unique meaning and must be processed by the UAC. The reason phrase is a more human friendly explanation, and need not be even processed by the UAC (the UAS need not even follow the recommended wording corresponding to a particular response code; it may choose to put the reason phrase in another human language, if it thinks that may be more convenient for the other side). Thus, for example, in the 200 OK response, the response code is 200, and OK is the reason phrase. In the 180 Ringing response, the response code is 180, and Ringing is the reason phrase.

All SIP responses fall into six categories. The response code's first digit indicates the category of a particular response. The categories of responses are:

- *100 class:* Provisional responses, given when a request has been received and is still being processed, but in the meantime the UAS wishes to let the UAC know the request has been received. Provisional responses are also known as informational responses. They are supposed to be sent if the processing in the server is expected to take over 200 ms to complete. Otherwise, the UAC may unnecessarily retransmit its request. 180 Ringing is an example of this class of response. If while locating the called party, a proxy is consulting a location service or doing something else that takes time, it could issue a 100 Trying response to stop retransmissions. In addition to 180 Ringing, the 183 Session Progress could be used to convey other information related to the call processing.
- *200 class:* Success cases, given when the request is successfully received and processed. 200 OK is an example of this class of response. It is currently the only 200 class response defined.

- *300 class:* Redirection, given when the UAC needs to take further action, such as to send to another address, to complete the request. 302 Temporary Redirect is an example of this class of response.

- *400 class:* Client error response, given when the request is invalid or cannot be satisfied at the server (but not because of a server error), as interpreted by the server, such as the case of a syntax error. An example of this response class is the 487 Request Terminated. Given in response to a CANCEL request subsequent to an INVITE, it is a case when the INVITE cannot be satisfied, but not because of a server error. The 400 Bad Request response is given when there is a syntax error in the request. Responses such as 401 Unauthorized, 403 Forbidden, and 404 Not Found, with which the user may be familiar from Web browsing (responses to HTTP requests), may also be returned.

- *500 class:* Server error response, given after the server fails to satisfy a request, although the request appears valid.

- *600 class:* Global failure response, meaning that the request cannot be satisfied by any server.

5.3.3   More on SIP Proxies

As mentioned earlier, the SIP proxies are network elements that consult a location service and forward requests. They pass responses back to the UAC that made the original request. For the duration of a transaction, the proxies involved in forwarding the original request will continue to be in the signaling path. The proxies do not have to remain in the signaling path for subsequent transactions in the dialog, but they can demand to be in the path, by using the Record-Route header, as explained earlier. The reader may have wondered how the responses know which proxies were involved in forwarding the request. Indeed, if the proxies simply forward the requests untouched, there is no way the UAS at the called party would know which proxies the request has traversed. Therefore, the proxy adds a *Via* header to the SIP header, with itself in the added header.

There are two kinds of SIP proxies, the *stateful* and the *stateless*. Stateful proxies keep the transaction state for all incoming requests. They also keep the transaction state for all requests generated by the proxy due to processing the incoming requests. In contrast, stateless proxies do not keep the transaction state for any transactions. They merely forward requests along, without making any associations between the two (since they discard information about packets after they finish processing and forwarding them).

Forking is a feature where a proxy can "try" multiple locations, perhaps because it obtains multiple addresses for a SIP UA from a location service. For example, if an INVITE comes in, a proxy could fork it to several locations with the hope that at least one is valid and appropriate for the SIP UA being searched for. A proxy must be stateful in order to support forking.

5.4 SDP, Parameter, and QoS Negotiations

SIP INVITE messages are typically sent with message bodies that describe various session-related parameters. SIP does not specify the format of its message bodies. Typically, these message bodies are formatted according to the SDP, which is fundamentally a format for conveying session description information, and information on real-time multimedia sessions in particular.

5.4.1 SDP Design Philosophy

SDP fits in very nicely with the piecewise, modular approach to the design of Internet protocols. The designers wisely do not tie it to a specific means of conveyance. Hence, SDP can be carried by a variety of protocols for a range of purposes. For example, it can support the advertising of multimedia conferences on the multicast backbone (mbone) using the session announcement protocol (SAP [5]). It can also be carried by e-mail with MIME extensions. Of course, it can be used by SIP as well.

We could think of SDP as a language for describing real-time multimedia sessions. However, the language can be communicated in various ways, including spoken, hand-written, and printed.

5.4.2 Using SDP with SIP

Interestingly, SDP itself "is not intended to support negotiation of session content or media encodings" [6]. However, the SIP session initiation model requires these capabilities. One of the attractions of the SIP model is that two parties who know very little about each other can negotiate a mutually agreeable set of encodings from the subset of coders that they have in common.

Since SDP describes session content and media encodings very well, what is needed to use it to support negotiation of session content and media encodings is simply a model for offering a description of a desired session from one party (possibly including lists of alternative encodings), and for the other party to answer with its own description of the desired session. This offer/answer model is usable as the basis of negotiations. In fact, such an offer/answer model does exist, and is referenced by the SIP RFC as the model to use with SIP [7].

When used with SIP as part of an offer/answer exchange, the SDP message is contained in the SIP message body, after the header. It is a valid SDP message as defined in RFC 2327 with a number of minor exceptions that are documented in RFC 3264 [6, 7]. For example, the initial offer contained in an INVITE message may look like this:

```
v=0
o=frodo 1234561234 1234561234 IN IP4 hobbit.shire.org
s=-
c=IN IP4 172.1.22.100
t=0 0
```

```
m=audio 7000 RTP/AVP 0
a=rtpmap:0 PCMU/8000
```

We examine the most significant lines. One line contains the connection information, which includes the IP address of the initiating party. For each media line in the SDP message, the media type (e.g., video or audio), port number, transport protocol, and media formats are specified. Typically, "RTP/AVP" is specified as the transport protocol, and then the media type is a profile number selecting from the profiles specified in RFC 1890 [8]. Note that the offer does not specify the port numbers of the RTP and RTCP packets it will *send*, but only at which it desires to receive packets.

5.5 SIP in Wireless Networks

SIP is not meant to be used only in wired networks—it can also be used in wireless networks. As such, it includes capabilities to handle mobility management [9], as will be explained in Chapter 6. It can work together with lower-layer mobility protocols like Mobile IP or GPRS, an example being the mobility management for IMS subscribers that we will see in Chapter 12, or SIP can handle mobility management by itself.

Additionally, bandwidth is a scarcer resource in wireless links than in wired links. Because they are text-based, SIP messages (including the text-based SDP messages) may be hundreds to thousands of bytes long. Clearly, text-based messages can be compressed for purposes of bandwidth efficiency. A general way to compress application protocol generated messages (e.g., SIP-generated messages) has been specified, where decompression can be performed by a universal decompressor virtual machine (UDVM) that can decompress messages that have been compressed by various popular compressors [10]. Using this approach, one way to compress messages is by using a static dictionary to translate well-known strings appearing in most SIP and SDP messages into more bandwidth-efficient sequences of bytes. Such a dictionary has also been specified [11]. Work is in progress on a signaling mechanism to signal when SIP compression is desired, by the use of a new parameter, "comp=sigcomp" in the SIP header [12].

5.6 Summary

While transport aspects of multimedia traffic over IP are a hot area of research, another equally important area is the signaling for initiating and maintaining multimedia sessions. The use of SIP signaling for setting up multimedia sessions is discussed in this chapter. We discuss the requirements for session initiation and how SIP meets the requirements. The fundamentals of SIP, including how other users are located, how and why registration is done, and the negotiation of session parameters, are explained. We examine SIP requests, replies, and proxies, and how SIP is implemented in a client/server manner, including as a text-based protocol. Since SIP is a text-based protocol, it may need to be compressed for use over bandwidth-scarce wireless links. References are made to some of the recent work on SIP compression.

References

[1]   Rosenberg, J., et al., "SIP: Session Initiation Protocol," RFC 3261, June 2002.

[2]   Johnston, A., et al., "Session Initiation Protocol (SIP) Basic Call Flow Examples," RFC 3665, December 2003.

[3]   Vemuri, A., and J. Peterson, "Session Initiation Protocol for Telephony (SIP-T): Context and Architectures," RFC 3372, September 2002.

[4]   Johnston, A., et al., "Session Initiation Protocol (SIP) Public Switched Telephone Network (PSTN) Call Flows," RFC 3666, December 2003.

[5]   Handley, M., C. Perkins, and E. Whelan, "Session Announcement Protocol," RFC 2974, October 2000.

[6]   Handley, M., and V. Jacobsen, "SDP: Session Description Protocol," RFC 2327, April 1998.

[7]   Rosenberg, J., and H. Schulzrinne, "An Offer/Answer Model with the Session Description Protocol (SDP)," RFC 3264, June 2002.

[8]   Schulzrinne, H., "RTP Profile for Audio and Video Conferences with Minimal Control," RFC 1890, January 1996.

[9]   Schulzrinne, H., and E. Wedlund, "Application-Layer Mobility Using SIP," *Mobile Computing and Communications Review (MCCR)*, Vol. 4, No. 3, July 2000.

[10] Price, R., et al., "Signaling Compression (SigComp)," RFC 3320, January 2003.

[11] Garcia-Martin, M., et al., "The Session Initiation Protocol (SIP) and Session Description Protocol (SDP) Static Dictionary for Signaling Compression (SigComp)," RFC 3485, February 2003.

[12] Camarillo, G., "Compressing the Session Initiation Protocol," draft-ietf-sip-compression-02.txt, October 2002, work in progress.

[13] ITU, "ISDN User-Network Interface Layer 3 Specification for Basic Call Control," Recommendation Q.931 (05/98), May 1998.

Appendix 5A   Notation

In this chapter, and in this book, we use the following notational conventions.

All SIP messages are spelled out completely in capital letters, whereas provisional responses only have their first letter capitalized. This follows the convention in the SIP RFC [1].

5A.1   ISUP Signaling

In circuit-switched systems like traditional phone systems, signaling is required to set up the circuit segments between the switches (the end-to-end circuit is the concatenation of the circuit segments). The main standard for signaling in modern[4] traditional telephony is SS7. SS7 uses a dedicated network for signaling, separate from the voice-bearing trunks between the switches. This signaling network is designed to be highly reliable and robust, with built-in redundancy of both the nodes and the links.

The SS7 standard includes a multilayered suite of protocols. The session-related signaling of interest is handled by the ISDN user part (ISUP) protocol [13] that uses either the underlying message transfer part level 3 (MTP3) or signaling connection control part (SCCP) for transport. Although the name might appear to imply that ISUP is only for ISDN calls, ISUP is in practice used for call setup and call tear-down in the telephone network, whether for ISDN or non-ISDN calls. In particular, it is used for the setup, management, and tear-down or release of trunk circuits (circuit segments of the end-to-end circuit) between switches. Since ISUP sets up trunk circuits between switches, it is not used when the calling and called parties use the same switch (since there is no interswitch segment in this case).

Figure 5A.1 illustrates the basic ISUP signaling. ISUP signaling is only between switches, so the ISUP messages (e.g., IAM, ACM) do not extend to the phones on

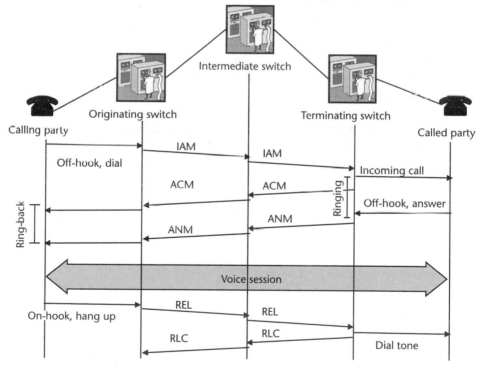

Figure 5A.1   ISUP signaling.

---

4.   There are older telephone switches that use other signaling schemes, some of which are of the in-band type, i.e., the same lines are used for signaling as for carrying the voice circuits. SS7 uses out-of-band signaling, where a separate, dedicated network is used for signaling.

either side. Instead, we indicate related events at the two phones by descriptive text, such as "Off-hook, dial" to indicate that somebody picks up a phone and dials a number. The initial address message (IAM) is generated by the closest switch to the calling party. We call this switch the *originating switch*. The IAM is analogous to the SIP INVITE message. It goes through one or more switches to reach the called party. The rest of the signaling will also go through the same set of switches, as the circuit connections are established between these switches. After the IAM arrives at the closest switch to the called party (a switch we will call the *destination switch*), the switch returns an address complete message (ACM) to indicate that addressing has been completed, which means a path has been found to the called party. As the ACM propagates back to the originating switch, one-way circuit establishments are completed, so that ring-back can be played from the destination switch to the calling party. The ring-back is played while the destination switch attempts to alert the called party. The ACM is analogous to the SIP 180 Ringing message.

When the called party answers the call, the destination switch sends an answer message (ANM) back to the originating switch. As the ANM propagates through the switches, the previously setup one-way circuits are modified to allow two-way traffic. The ANM is analogous to the SIP 200 OK message. Because of reliability mechanisms in SS7, there is no need for something analogous to a SIP ACK message.

Despite the similarities to the SIP signaling, there are some interesting differences between the sessions set up by ISUP signaling and those set up by SIP. In this case, actual voice circuits are set up between the switches, where dedicated resources are set aside for the 64-Kbps PCM-encoded voice. Also, the voice path will go through the same switches that the ISUP signaling goes through. With SIP on the other hand, no actual voice circuits are set up.[5] Instead, UDP ports on both sides, session parameters, and media encodings are agreed upon between the two ends, and packetized voice can flow directly between them.

---

5.   When we look at QoS for IP networks, we will see that some of the techniques involve resource reservation (e.g., bandwidth reservation) so that the quality can be somewhat akin to that of actual circuits.

CHAPTER 6
Mobility Management

Wireless is about freedom of movement. Managing the mobility of users is therefore one of the key technical challenges of wireless IP research. It would seem that a user should be able obtain some IP services no matter where the user is located. For example, you would like to be able to take your laptop with you to places other than where you normally use it, and still be able to obtain some IP services, even if not as rich a set of services as you might be used to.

One can divide the problem into two parts. First, there is the issue of selecting an appropriate point of attachment to the network and establishing the link through that attachment point. Second, after the low-level link has been established, higher-level services (such as network services like IP routing, transport layer services like TCP connections, and higher-layer concepts like SIP-based multimedia sessions) need to be made accessible at the new attachment point. When selecting an appropriate point of attachment to the network, various factors are typically considered, like signal strength of the alternative choices (whether they be base stations of cellular systems, or access points of WLAN). Many algorithms have been proposed, but this selection and the link establishment are mostly radio issues below the network layer and beyond the scope of this book [1, 2], although we have briefly introduced some of the ideas like soft handoff in Chapter 3. In this chapter, we instead focus more on the second subproblem, describing protocols such as Mobile IP to make IP routing work in the presence of mobility. The main ideas include registration of the location of the mobile node and mechanisms to route traffic to the correct location. We note, though, that recently cross-layer techniques are being studied, especially in the context of reducing handoff latency. We discuss these ideas in Section 6.4.

We might expect that switching between network attachment points would need to involve network-layer protocols. However, in some cases, link-layer protocols can handle some "local mobility" transparently to the higher layers, as we will discuss in Section 6.2.3. There are cases where existing services are being utilized at a previous attachment point, and these services need to be smoothly transferred to the new attachment points. This is the handoff problem, because connectivity and services need to be carefully and properly handed off between attachment points. In Section 6.2.1, we will contrast this with the less-challenging (but still interesting and important) problem of location management, where the network just needs to ensure the mobile node can be located in order to utilize services, such as to start a new SIP session.

6.1   A Network-Level Solution: Mobile IP

One of the earliest mobility management protocols proposed was a protocol called *Mobile IP*. Despite some drawbacks, Mobile IP has stood the test of time and is well established today. It is a standard feature included in routers and is deployed in cdma2000 networks.

6.1.1   The Problem Addressed

We take postal mail delivery as an analogy. Most people have a home address that uniquely identifies a location, such as a residence or a post office box. When they move and obtain a new address in the new location, they would like their mail to reach them at the new location.

The problem addressed by Mobile IP is similar. Most machines on the Internet have an IP address that uniquely identifies a location, in the sense of a certain subnet region of the whole network. When machines move and obtain a new address in the new location, they would like IP packets addressed to them to reach them at the new location. We refer to such nodes that move as mobile hosts (MH). Note that Mobile IP is not specified only for wireless nodes, but can be used by nonwireless nodes too. For example, a laptop that is brought around to different locations could take advantage of the services provided by Mobile IP. Nevertheless, Mobile IP is most useful for wireless nodes.

6.1.2   Mobile IP

We first continue the postal-mail analogy to provide an overview of the Mobile IP solution. We then revisit Mobile IP in more detail and follow with a qualitative assessment.

6.1.2.1   Overview by Analogy

Suppose Jane moves from her apartment in New York to a house in Singapore. People who send her mail but who do not know her new address will send the mail to her New York address. Therefore, Jane sets up mail forwarding with her local post office in New York to forward her mail to Singapore. Thus, whenever a piece of mail for Jane arrives at the post office in New York, a postal worker places the unopened envelope inside another, slightly bigger envelope. The destination address on the new envelope is the address in Singapore. When the piece of mail arrives at the Singapore address, Jane's secretary opens it, takes the original envelope out of the outer envelope, and hands it to Jane. In the other direction, on the other hand, mail from Jane to her friends can go directly to them from Singapore without needing to go through New York.

Mobile IP works in an analogous way; we will overview the Mobile IP[1] solution here to highlight its similarities to the postal-mail analogy, and then provide more

---

1.   There is also a version of Mobile IP for IPv6. In this chapter, we discuss the original Mobile IP protocol and leave Mobile IP for IPv6 to Chapter 9 when we consider IPv6.

details in Section 6.1.2.2. Suppose an MH moves from its home network to a foreign network, as shown in Figure 6.1 (the analogous movement of Jane from New
York to Singapore is shown in Figure 6.2). Other machines, also known as correspondent hosts (CHs), that are sending packets to the MH will direct the packets to
its home network address. So, the MH sets up packet forwarding from its home network to a new address in the foreign network. The process of setting up the packet
forwarding is known as *registration*. A home agent (HA) in the home network
agrees to intercept packets addressed to the home address of the MH, and to forward such packets to the foreign network. The packets are forwarded by encapsulating them in new packets addressed to the foreign network address (see Figure 6.3
for an example of one such encapsulation scheme, namely, IP-in-IP encapsulation).
When the packet arrives at the foreign network, another mobility agent, a foreign
agent (FA), unencapsulates (unencapsulation is also known as decapsulation) the
original packet and hands it to the MH. In the other direction, packets from the MH
can go directly to CHs without needing to go through the home network of the MH.
This operation of mobile IP is illustrated in Figure 6.4.

Figure 6.1   Moving to a different network.

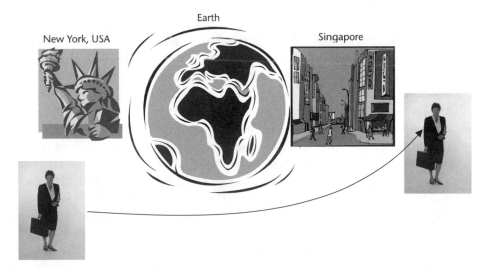

Figure 6.2   Analogy for moving to a different network.

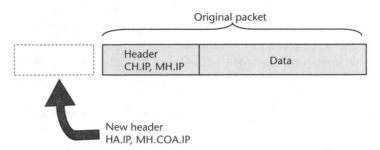

Figure 6.3 IP-in-IP encapsulation.

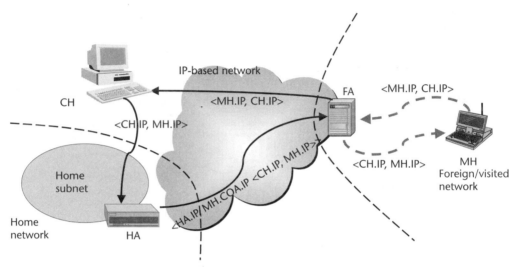

Figure 6.4 The Mobile IP solution, with FA.

6.1.2.2 A More Detailed Look

As discussed in Chapter 2, the Internet uses a hierarchical addressing and routing scheme, instead of putting specific routing information on each individual valid IP address in every router (an alternative that is clearly not scalable). The hierarchical addressing and routing scheme is based on a way of thinking of IP addresses as being concatenations of a network prefix with a host address, and distribution of chunks of contiguous IP addresses to organizations, where these IP addresses would have the same network prefix. This allows the very convenient trick of routing to a network—each router in the Internet backbone need only have one entry in its routing table for that entire network. All packets addressed to an IP address with that network prefix are then conveniently routed using that one entry. However, this notion has the unfortunate side effect of tying IP addresses to locations—all packets destined to an IP address will be routed to a particular network, due to the hierarchical routing scheme. We call this network the *home network* for the IP address. If that IP address is the "permanent" address of an MH, we can speak synonymously of the home network for that IP address as the home network of the MH as well. We call that IP address the home address, to differentiate it from any temporary address the MH may pick up in a foreign network. A foreign network would be any other network an MH could find itself in, that being where packets routed to its home address would normally not be sent.

The problem that Mobile IP addresses is that of packets arriving at an MH's home network while it is *roaming*, away from its home network in a foreign network. The challenge is how to arrange for these packets to be delivered to the MH in the foreign network, while constraining the solution to, first, not require changes to the existing Internet routers and the addressing/routing scheme and, second, not require changes in the CH. These constraints are very important because otherwise implementation of the solution would be quite impractical. With the constraints in place, the solution could be gradually implemented without affecting existing functionality, as has turned out to be the case. Mobile IP can be thought of as an overlay solution over the existing IP, rather than a redesign or extension of IP that might not satisfy the two design constraints.

Since the solution could not require changes in CHs, or changes in the existing Internet addressing/routing scheme, it has to deal with packets for the MH arriving at its home network regardless of whether the MH is roaming or not. When the MH is at home, Mobile IP does not interfere. However, when it is roaming somewhere, a mobility agent in the home network, the HA, intercepts the packets on behalf of the MH. It may need to use mechanisms like "proxy ARPing," depending on the physical- and link-layer technologies used in the home network [3]. This interception of packets is analogous to a friend picking up your bags from an airport baggage claim belt when you are unavailable to do it yourself. Otherwise, the packets would be unclaimed and lost. The next step is to deliver each intercepted packet to the MH at the foreign network. The HA uses the foreign network address for this purpose, and the foreign address is known as the *care-of address* (COA). This address is obtained through Mobile IP registration (we will discuss registration further in a few paragraphs). Given the constraints we mentioned, the solution needs to make use of the existing IP routing scheme to effect this delivery. It might appear that the most convenient way to do this is to replace the IP header of the original IP packet with a new IP header using the HA address as the source address and the COA as the destination address. However, this solution is problematic because the original IP header is lost and cannot be completely reconstructed. Therefore, the HA instead could put the entire original IP packet into a new packet with a new header using the HA address as the source address and the COA as the destination address. Indeed, this will work, and is known, not surprisingly, as IP-in-IP encapsulation: the original IP packet is encapsulated in a new IP packet. Other encapsulation schemes are also possible, the only requirement being that the original packet can be completely reconstructed at the destination. *Minimal encapsulation* adds only a minimal 8 to 12 bytes to the original packet.

The encapsulated packet will be routed to the foreign network by normal IP routing. There are two possibilities for the COA: It can either be the address of a FA in the foreign network, or it can be a colocated COA that the MH has acquired in the foreign network through other means [e.g., using Dynamic Host Configuration Protocol (DHCP)]. If the COA is a colocated COA, the MH will perform the unencapsulation itself, and the operation of Mobile IP in this case is shown in Figure 6.5 (which can be compared with the case with FAs, as seen already in Figure 6.4). If the COA is instead the address of an FA, the FA will unencapsulate the packet, removing the outer header and taking the original, inner packet out. It would then deliver the packet to the MH. Note that this delivery should be a link-layer delivery, meaning that IP routing cannot be used to deliver the packet, or else the packet would be routed back to the home network of the MH (since the destination address is the home address of the MH). Similarly, in the postal-forwarding example, Jane's secretary should hand the original envelope to Jane in person, rather than dropping it in a mailbox (here the analogy is somewhat imperfect—in the postal case, the letter would not be delivered back to New York, because a used stamp is attached).

So we have seen that Mobile IP basically operates by tunneling packets from a MH's home network to the foreign network where the MH is currently located; the HA is the forwarding agent. But how does the forwarding get set up in the first place, and how is it updated when the MH moves so that the HA can keep forwarding to the current location of the MH? The answer is the registration process. We first consider the case when FAs are used, as shown in Figure 6.6. When the MH moves into a new network, it may detect that through a variety of means, including internal messages from L2, or receiving an agent advertisement message from a new FA (message 2 in the diagram). The MH could also optionally broadcast an agent solicitation message (message 1 in the diagram) to hasten the receipt of an agent advertisement message. The MH learns the FA IP address through the agent advertisement. It sends a Mobile IP registration request message to its HA (message 4), relayed through the FA (message 3). The registration request contains the FA IP address as the COA of the MH. If all is in order, the HA returns a positive response (messages 5 and 6). Why

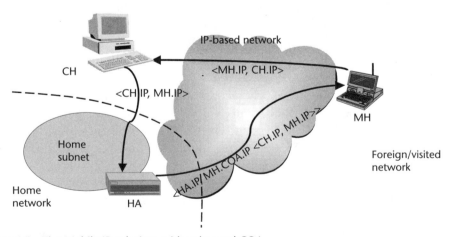

Figure 6.5 The Mobile IP solution, with colocated COA.

might all not be in order? Perhaps because of an authentication failure (the registration request contains authentication information from the MH), as elaborated in Chapter 8. Now, in the case of colocated COAs, as shown in Figure 6.7, the MH first obtains a COA in the foreign network through some other means, such as through DHCP. It then registers directly with its HA (messages 2 and 3).

Although we at first focused on the problem of an MH moving from its home network to a foreign network, Mobile IP is equally applicable in the case of moving from one foreign network to another. The signaling protocol for the registration is the same whether the MH was last at its home network or a foreign network. Similarly, the operational details after the registration are the same: Packets are tunneled to the appropriate foreign network. Of course, there may be some software differences in the handling of the two cases in the HA and the MH. For example, in the

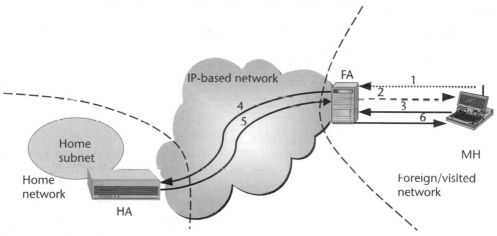

Figure 6.6   Mobile IP registration (through FA).

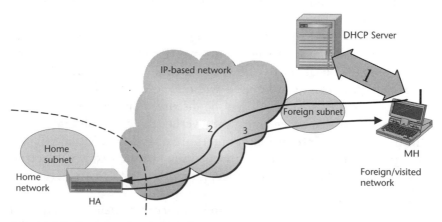

Figure 6.7   Mobile IP registration (with colocated COA).

HA, in the former situation it would have to "turn on" packet forwarding for the MH, creating an entry in an internal table for the binding of its home address to its COA. In the move between foreign networks, it may be more of a question of updating the existing entry in that binding table. Nevertheless, the communication protocols for handling the home-to-foreign and foreign-to-foreign moves, as specified by Mobile IP, are the same in both cases. Hence, typically, and correctly, the two cases are not distinguished in theoretical discussions of Mobile IP, but are distinguished in discussions on implementation.

6.1.2.3   Assessment of Mobile IP

The biggest highlight of the Mobile IP solution is that CHs need not be modified to support mobility. All the functionality to support mobility is in the home network of the MH (in the HA), in the MH, and possibly in the foreign networks to which it roams (if FAs are used). Any ordinary Internet host can be a CH. Mobile IP does not require that a CH have additional functionality to support MH mobility, or that it needs to be aware even that it is talking to a mobile node. This allows practical, gradual deployment of Mobile IP; even if at first only a small number of hosts support Mobile IP, the protocol allows them to communicate with the millions of unchanged potential CHs out in the network.

In keeping with modular design principles, even in hosts like the MH where changes in functionality are required, Mobile IP minimizes the impact of mobility on the transport layer and above. This notion has often been expressed as that Mobile IP is *transparent* to higher layers. It is right in the sense that this means the higher layers do not "see" the mobility because they are only aware of the home address of the MH and the address of the CH, and not aware of the COA and tunneling that goes on beneath. However, care should be taken not to stretch the notion of transparency to mean that the upper layers are always completely oblivious to the mobility handling beneath. In fact, delay-sensitive applications may be impacted by handoff latency and may need to adjust buffer sizes, as well as other elements, to deal with it.

Mobile IP can be thought of as an *overlay* on an IP-based network. It does not change IP itself, but makes use of normal IP routing mechanisms, with some crafty fashioning of IP headers, to route packets to the desired locations. This is why all the intermediate routers can be used without needing to be aware of Mobile IP (one exception is firewalls, which will be discussed shortly), which is an advantage of Mobile IP. On the other hand, the overlay design results in some disadvantages. Since the CH and intermediate routers are unaware of Mobile IP, Mobile IP resorts to a triangular routing scheme through the HA. Sometimes, this triangular routing can be very inefficient, for instance, if the MH and CH are very close to each other while the HA is on the other side of the world.

While the triangular routing can be a very inefficient use of network resources at times, a more serious problem is handoff latency. Handoff latency refers to the time it takes to complete a handoff, during which packets can be lost because they are not yet routed to the new point of attachment. Many enhancements and modifications to Mobile IP have been proposed to address this issue, as we will see in this chapter. Some of these enhancements and modifications help address some of the other weaknesses of Mobile IP.

These other weaknesses of Mobile IP include:

- *A single point of failure:* If the HA fails, all the MHs it supports become unreachable. Since the failure of only a single node can bring such grief, this node can be described as a "single point of failure," a term from the literature of system reliability and robustness.
- *Problems with firewall traversal:* A firewall in the foreign network may perform *egress filtering*, whereby it drops packets being sent out of the foreign network with a source address that is not part of the foreign network. Since packets from the MH are sent with the home address as source address, these will be dropped.
- *Encapsulation overhead:* Tunneling from the home to foreign network involves encapsulation that adds at least 8 to 12 bytes of overhead; this gets significant as packets get smaller.
- *Signaling overhead:* Each time an MH moves into a new foreign network, and also periodically (re-registration), it needs to register with its HA. This overhead is particularly wasteful if the MH is idle for a long period of time but moving frequently during that time.

A large body of work deals with the handoff latency problem. This work will be discussed in Section 6.4. As side effects, signaling and encapsulation overheads are also reduced by some of the proposed modifications and enhancements to Mobile IP that would be presented. Meanwhile, an attempt to deal with the triangular routing problem will be explained in Section 6.1.2.4. Firewall traversal issues can be handled through solutions, such as "reverse tunneling," that will not be covered in this book.

6.1.2.4   Route Optimization

To reduce the occurrence of triangular routing, it has been proposed that MHs send a binding update message to all known CHs whenever handoff occurs [4]. The binding update will provide CHs with the new COA of the MH. The CHs then have to option to tunnel packets directly to the foreign network (using the COA), without needing to go through the home network. Regular Mobile IP registration continues, so if there are packets from any CHs that do not know the new COA of the MH, the HA can tunnel them as usual. One example is new CHs wanting to initiate a session, that have not received a binding update from the MH. Another example is CHs that do not implement the additional functionality required in CHs for route optimization to work. In both these cases, packets will be forwarded by the HA as in traditional Mobile IP. Route optimization is shown in Figure 6.8.

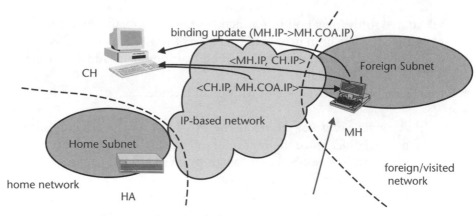

- Removes triangular routing
- CH to MH
  - In this illustration, we assume colocated COA is used
  - CH sends packet to MH.COA.IP directly
- MH to CH
  - Normal IP routing from foreign network

Figure 6.8   Mobile IP with route optimization.

The route optimization proposals for Mobile IP have traditionally been accompanied by a proposal for a scheme to reduce the number of dropped packets during handoff. We will discuss this scheme in Section 6.4.6.

6.2   Mobility Concepts

6.2.1   Location Management and Handoffs

One of the earliest distinctions between mobility problems is the distinction between location management and handoff management. Even back with the first generation of cellular systems, people noticed that it was one thing to keep track of subscribers while they were not in a call and another to keep track of subscribers while they were in a call.

The main reason for keeping track of idle subscribers (subscribers who are not in a call) is to be able to initiate a call with them, for an incoming call. Tracking a mobile subscriber can be performed at different levels of precision. As explained in Chapter 3, it may be better not to keep track of idle subscribers at the level of a cell, but rather at the level of a location area covering multiple cells, in order to save on radio resources. To contact these subscribers, for instance, to respond to an incoming call, paging is then required. Compared with handoff, updating of the location is not highly time critical, because a slight delay in updating of the location may not be noticeable, given the delay that people expect in call setup.

The handoff management problem involves keeping track of subscribers who are in an active call, as they move from BS to BS. Compared with location management, handoff management requires more careful, timely treatment, because a session is active during the handoff, and even small delays can be irritating to the subscribers.

The distinction between the two situations is not merely academic but is useful for deriving two sets of requirements for the two situations. In general, location management is easier, has looser constraints, and consumes fewer resources than handoff management. Therefore, wherever possible, it is better to perform location management than handoff management. However, we need to take care that we do handoff management rather than location management in situations where it is required, to avoid unnecessary increases in session interruption times during handoffs.

The reader may, in considering the last few sentences, wonder whether there really is an issue deciding between location management and handoff management or if it is not a clear-cut decision, based on whether a session is ongoing. Indeed, for voice communications, the answer is clear. However, things are not so simple when data communication is considered. How do we decide when there is an active session, so that handoff is needed, and when there is not, so that location management is needed? We will later discuss how 802.11-based wireless LANs and GPRS handle this issue.

6.2.2 Types of Mobility

We have seen what Mobile IP does, in Section 6.1.2. Mobile IP is one technique for handling *terminal mobility*. By this, we mean that it helps maintain network services when terminals move. Terminal mobility can be either of the location management or handoff management type. Another type of mobility is *personal mobility*, where services are maintained when "persons" move. We can think of person in this context as the user of a machine, whereas terminal mobility is about terminals moving, as shown in Figure 6.9. If a person moves from machine to machine and the network keeps track of the movement and can find the person to initiate sessions, we say that personal mobility is supported.

Personal mobility can be thought of as location management for persons; more precisely, it is about keeping track of the *changes in association* of persons with terminals. In a GSM system, for example, the SIM card can be moved from phone to phone, and the system properly tracks the card's movement (rather than the phone's). This is a personal mobility solution. What about personal mobility in a handoff situation? Both the terminal and the person move from one AP to another (or one BS to another). However, this situation is typically considered to be terminal mobility rather than personal mobility, because the person is associated with the same terminal before and after the handoff. Personal mobility can be accomplished by registration of location whenever a person moves, as seen in Figure 6.10, or by using multiple simultaneous registrations at multiple locations so a person remains reachable after movement, as seen in Figure 6.11.

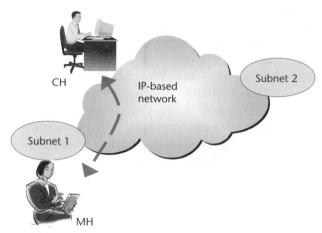

Figure 6.9   Communications is not just between terminals; there are persons and sessions.

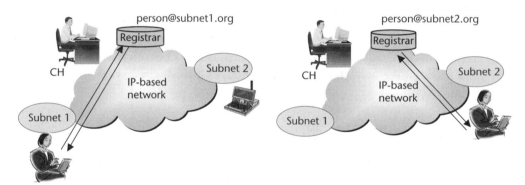

Figure 6.10   Personal mobility.

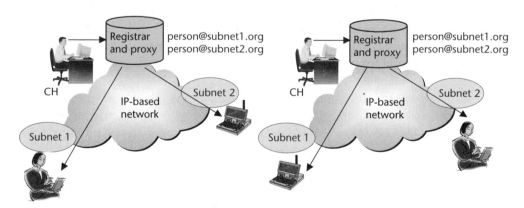

Figure 6.11   Personal mobility through registration at multiple locations.

Terminals are used by persons, and the applications used by these people often work with communication sessions (e.g., analog voice sessions, voice or video over

IP sessions, and TCP sessions). In addition to terminal mobility and personal mobility, then, one would expect there to be a notion of *session mobility* as well. Indeed, there is, and it has to do with the ability to transfer an ongoing session from one device (terminal) to another. For example, a subscriber might be using her cell phone to speak with a friend, and while they are talking she arrives at her office and wishes to transfer the call to her office phone. With today's cell phone systems, it may not be possible to do such a transfer, so what might happen is that she hangs up and redials her friend from the office phone. However, this should not preclude adding session mobility support in future multimedia over IP communications protocols. We will later see how SIP explicitly incorporates session mobility capabilities, as shown in Figure 6.12. (The messages shown in the figure are particular to a SIP handling of session mobility, to be discussed in Section 6.3.1.2.)

Finally, there is the concept of *service mobility*. It is about experiencing a service environment similar to the service environment one is used to at home, even while roaming. The service environment includes the "look and feel" of the user interface, access to the same sets of services as one can obtain at home, and access to those services in the same manner as one is used to. This involves more than just rerouting of packets. Going back to the analogy involving Jane moving from New York to Singapore, perhaps she finds it hard to immediately adjust to life in another culture, so she looks for an expatriate community in which to live, a place where there are grocery stores selling goods she is used to, close to an international school. All this is

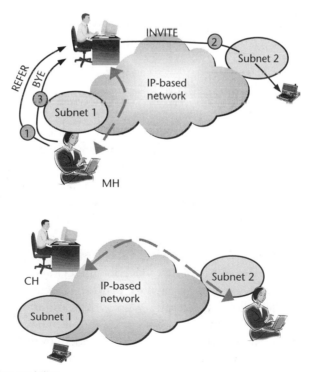

Figure 6.12   Session mobility.

in addition to arranging for the mail forwarding from the post office in New York. Service mobility is a similar idea.

It should be noted that terminal mobility, personal mobility, session mobility, and service mobility are concepts associated with different but related problems. Solutions such as Mobile IP and SIP mobility may provide one or more of these types of mobility, and may be applicable for handoff and location management.

6.2.3 Layer 2 Versus Layer 3 Mobility

The next concept we explore is that some aspects of mobility may be handled at the second layer of the OSI protocol stack (the link layer) and some other aspects may be handled at the third (the network layer). We call the handling of mobility at the link layer, Layer 2 mobility (or L2 mobility for short), whereas the handling of mobility at the network layer we call Layer 3 mobility (L3 mobility). Traditionally, the characteristic function of the link layer is to transmit information across a link (one "network" hop) [5]:

- It detects data corruption;
- It coordinates the use of shared media (typically broadcast medium like Ethernet LAN, but could be single-user too, such as voice modem dial-up);
- It provides link-layer addressing when multiple nodes are reachable (Ethernet MAC addressing is a good example).

On the other hand, the network layer enables any pair of nodes in the network to communicate, even over multiple hops:

- It calculates routes;
- It fragments and reassembles packets;
- It provides some congestion control.

Applying these ideas to mobility management, then, we would expect Layer 2 mobility to address mobility within a LAN and Layer 3 mobility to address all other mobility. For example, Mobile IP, a Layer 3 solution, would not kick in for mobility within a LAN, but only when the movement is between different LANs. Examples of traditional Layer 2 mobility are shown in Figure 6.13. Wireline examples include the Ethernet LAN, and voice modem dial-up, with a protocol like Point-to-Point Protocol (PPP) running over the physical links. Wireless examples include traditional wireless data services like Mobitex, ARDIS, and cellular digital packet data (CDPD). As these are low-rate, wide-area wireless data services, they are not suitable for high-rate wireless data expected by Web surfers and multimedia communications.

However, more recently concepts of L2 mobility have been extended so that significant mobility management occurs at L2, transparent to the IP layer, as shown in Figure 6.14. Mobility management internal to an 802.11 ESS is a great example of extended L2 mobility, and will be discussed in Section 6.4.3. GPRS is a special

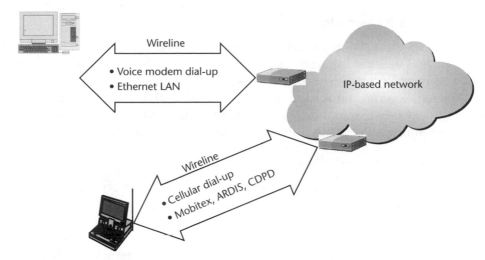

Figure 6.13   Traditional forms of L2 "last-hop" link.

overlay upon, and extension of, cellular networks to better support packet data services. It also provides extended L2 mobility support, and will be discussed in Chapter 11.

6.3   Alternative IP Mobility Schemes

Since IP is a network layer protocol, it may appear natural to handle mobility at the network layer. Indeed, this is how Mobile IP does it. However, there are alternatives, towards which we now turn our attention. We note that elements of SIP mobility (to be introduced next) are finding their way into specifications being developed in 3GPP and 3GPP2 (see Chapter 12 for an example).

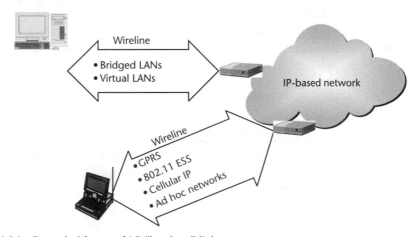

Figure 6.14   Extended forms of L2 "last-hop" link.

6.3.1 SIP-Based Mobility

SIP, as discussed in Chapter 5, is used for initiation and management of sessions such as multimedia-over-IP sessions. SIP can also be used to support mobility management. It is powerful enough to support not just terminal mobility, but also personal, session, and service mobility [6]. SIP mobility management can be divided into presession mobility management and midsession mobility management, depending upon whether the subscriber is in a SIP-managed session or not.

6.3.1.1 Presession Mobility

Presession mobility (also known as out-of-session mobility) capabilities provided by SIP include the following:

- Ability to register a SIP user's presence in his or her current location;
- Ability to register and map one SIP address to several devices (known as forking);
- Ability to register and map several SIP addresses to one location.

These capabilities allow SIP to support presession terminal mobility, and also personal mobility and service mobility. Presession terminal mobility occurs when a terminal moves and obtains a new IP address. SIP registration is used to update the bindings at the SIP location service (the SIP application in the terminal sends a REGISTER message to the SIP registrar, so that the binding can include the latest IP address). Personal mobility can also be handled using SIP registration, since a SIP REGISTER message can indicate that user@olddomain be registered at user@newdomain. Thus, bindings of SIP address to SIP address support personal mobility (since the linking is at the user or personal level), whereas bindings of SIP address to IP address support terminal mobility (since the linking is to a particular terminal's IP address). Another way that personal mobility is supported by SIP is by allowing one SIP address to be registered to (and thus have multiple bindings to) multiple devices. Thus, a user can move between devices without needing to register his or her new location each time.

It is currently less clear how service mobility might be supported by SIP. The obvious candidate is to enhance the SIP registration procedure to handle service- and subscription-related parameters.

6.3.1.2 Midsession Mobility

Midsession mobility is closely tied to handoff, especially in the case of midsession terminal mobility. However, midsession mobility also includes session mobility, where a session is transferred in midsession from one device to another.

SIP provides the following mechanisms to support midsession mobility:

- Ability to re-INVITE a SIP end-point when there is a change of IP address;
- Ability to transfer a session from one terminal to another by referring it to the other terminal.

Handoffs may or may not result in a change of IP address. Examples of no change of IP address include L2 mobility, such as movement between APs in the same wireless LAN ESS, and movement between BSs under the same cellular IP gateway (further details can be found in Section 6.4). If there is an SIP-managed session between the MH and a CH, it clearly will not be affected by handoffs where there is no change of IP address. However, if the IP address changes, the CH needs to be informed of the new IP address, or it would continue sending packets to the old IP address. The way a MH would handle this is to send a re-INVITE to each CH with which it has current SIP-managed sessions. A re-INVITE is basically an INVITE message that is sent to the CH with the same call identifier as the original INVITE (thus informing the CH that this corresponds to the earlier session, and is not a new INVITE for a new session). The new IP address is placed in two locations: (1) in the Contact field of the re-INVITE, so SIP signaling will go to the new IP address; and (2) in the SDP portion of the re-INVITE, so that the RTP streams get redirected to the new IP address as well.

SIP can also provide session mobility, through one of at least two ways of handling session transfer. An ongoing session can be transferred from one device to another using either *third-party call control* or the *REFER method*. The REFER method is a new SIP method whose use is more elegant than the third-party call control solution (for session transfer; third-party call control is still a useful concept for other scenarios). Using the REFER method, imagine Jane is speaking to John (john@somewhere) and wants to transfer from her mobile phone, where she is jane@mobile, to her regular phone, where she is jane@notmobile. The first message sent is a REFER jane@notmobile message from jane@mobile to john@somewhere, where the "Referred-By" field contains jane@mobile. John@somewhere then sends a regular INVITE message to jane@notmobile, where the "Referred-By" field contains jane@mobile.

6.3.2   Transport-Layer Approaches

We have seen how it is more appropriate to handle mobility management on the global scale at the network layer than at the link layer. But what about higher in the protocol stack, such as the transport layer? In fact, a number of mobility management schemes *have* been proposed that operate at the transport layer. They share the following characteristics: (1) session specificity, meaning specific to the context of a session, such as a TCP session, (2) modifications to existing transport layer protocols, and (3) implementation that is not wide.

TCP migration and MSOCKS are transport layer approaches to mobility management that modify TCP to provide handoff support [7, 8]. MSOCKS uses a special proxy to support mobility. A TCP connection between host A and MH B is really two TCP connections, one between A and the proxy, and the other between B and the proxy. The proxy makes the two TCP connections appear to be one connection between A and B, with a technique called *TCP splicing*. It requires special software both on the proxy and on the MH. It intercepts TCP connect() requests from the application on B, and exchanges information with the proxy (over a normal TCP connection between B and the proxy) to support the splicing and the mobility support. When B moves, the TCP connection between B and the proxy is reconnected. However, the software on B hides the fact from the application, while the proxy splices the new connection to the proxy connection to A so that A is unaware of the mobility. While there are other transport-layer mobility protocols that use proxies, MSOCKS has some uniquely attractive features. It does not require the CH to be modified (i.e., the CH can have an ordinary TCP stack). Also, it does not violate the end-to-end semantics of TCP, because the proxy performs TCP splicing. Thus it appears like an ordinary TCP connection to the applications on both sides, and they are unaware of the reconnections when the MH moves and obtains a new IP address.

TCP Migration is a more recently proposed transport layer mobility management scheme that does not require proxy support. However, both ends of the TCP connection need to be slightly modified to support a new *TCP Migrate* option (alternatively, the proposers of TCP migration have suggested that a proxy *can* be used to provide for communications with a nonmobile host that does not support the TCP Migrate option). The proposers of TCP migration consider this an acceptable requirement, since they follow the end-to-end design principle [9]. When an MH moves and tries to reconnect with the other host, the TCP Migrate option allows the logical linking of the old and new connections. An advantage of TCP migration over schemes like MSOCKS is that the initiation of the connection could come from either side, not just the MH. Although in many cases, communications would be client-server based and initiated by the MH, this is not always true, so this is a valuable feature. Note that the TCP Migrate option supports handoff and not location management. In order to allow a potential CH to find an MH to initiate a TCP session, there must be a way for the CH to obtain the current address of the MH. Secure dynamic DNS upgrades have been suggested to solve this location management problem. Since the use of dynamic DNS is not specific just to TCP migration, but should be useful for other transport layer schemes as well, we discuss it as a separate topic in the next section.

Transport-layer schemes provide an interesting alternative approach to mobility management for IP networks. However, they have not gathered the same kind of momentum in the industry as Mobile IP-based and SIP-based solutions have gathered. Part of the reason is the drawbacks of these schemes. For example, they only support TCP sessions, so traffic carried by UDP is not properly handled by these schemes. Also, if there are many TCP ongoing sessions when an MH moves, each has to be handled separately. With Mobile IP, on the other hand, the same amount of signaling is involved regardless of how many TCP sessions are connected.

6.3.3 Dynamic DNS

Versions of TCP modified to support mobility, as described in Section 6.3.2, handle handoff only and not location management. A complete solution would provide location management capabilities, so that when MH moves while not currently engaged in any TCP session with CH, the location is tracked to enable a later TCP session initiation with the MH. The natural place to add this capability is DNS.

In fact, dynamic DNS capabilities have been proposed [10], and its use in location management (as part of the TCP migration solution) has also been suggested [6, 7]. However, there are several issues with dynamic DNS for location management. First, there should be a way to ensure that potential CHs do indeed query DNS whenever they want to start a session with the MH, and that they do not sometimes use cached, possibly stale IP addresses. This is indeed done by the great majority of applications, although local caching may still be a problem in some cases. Second, even if CHs query DNS accordingly, there is an issue with DNS servers themselves using cached information, and thus providing outdated IP addresses. Snoeren addresses this issue by proposing that the TTL field for the DNS entry A-record (i.e., for the IP address) for the MH be set to 0 to prevent caching of potentially outdated IP addresses [7]. Third, security is critical—if a malicious host successfully gets the wrong IP address for a target host into DNS, the target host could be shut out from receiving future communications.

Of course, dynamic DNS by itself is not a complete mobility management solution either, since it supports only location management and not handoff. No CH would keep querying a DNS server frequently enough that we could claim that dynamic DNS supports handoff!

6.4 Micromobility and Fast Handoff

One of the major headaches in using Mobile IP is long handoff latency. Also, the global update signaling overhead could be a problem as the use of Mobile IP increases. Among the most prominent solutions, based on reducing the network latency of updates, are the micromobility schemes. Conceptually, these schemes break up the roaming regions into micromobility zones (although they may not actually use the term "micromobility zones"). The key characteristics of micromobility zones are: (1) they are larger than "normal" coverage areas of wireless IP subnets (for example, a region where there might have been a number of subnets each covered by a FA, might be replaced by one micromobility zone, as shown in Figure 6.15), (2) Mobile IP global registration is not needed for movement within the micromobility zone; only local updates are needed, and (3) Mobile IP global registration is needed only for movement *between* micromobility zones. The idea is that most of the handoffs of the MH would occur *within* micromobility zones (with only a few between micromobility zones). Only local updates would be needed for these, and not the normal Mobile IP registration (which is global in scope). Hence, the expected handoff latency would be very low. Global signaling overhead is also significantly reduced.

We note that because of certain shared characteristics, Layer 2 mobility schemes often can be thought of as micromobility schemes. However, micromobility schemes need not be implemented in Layer 2. Hierarchical Mobile IP is a Layer 3

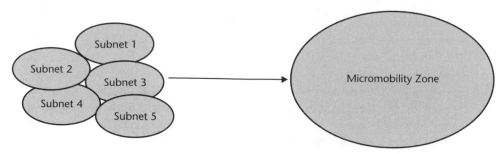

Figure 6.15   Replacing multiple FAs in multiple subnets with micromobility zone.

approach to micromobility, whereas host-based routing can be implemented in Layer 2, or in a pseudo-layer 2 manner in Layer 3 (i.e., through manipulating Layer 3 routing tables in a way typical of Layer 2 mobility). The 802.11 WLAN micromobility is purely a Layer 2 scheme. GPRS micromobility is Layer 2 in style, although it is intriguing in the way it uses IP at multiple layers.

6.4.1   Hierarchical Mobile IP

One way that micromobility could be implemented is with a two-level hierarchy of foreign agents. For example, in Mobile IP with regional registration [11] (see Figure 6.16), a regional foreign agent (RFA) replaces each FA, and groups of RFAs are "under" a gateway foreign agent (GFA). Unlike regular Mobile IP, the HA of each MH will tunnel packets to the corresponding GFA. The GFA un-encapsulates the packets and tunnels them to the appropriate RFA. Notice that this arrangement means that the HA needs only to keep track of the GFA, and need not be informed when the MH moves between RFAs under the same GFA. As far as the HA is

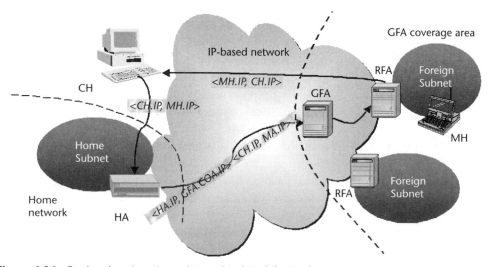

Figure 6.16   Regional registration, a hierarchical Mobile IP scheme.

concerned, it needs only to know how to tunnel the packets to the correct GFA. Thus, location information is hierarchically distributed, with the HA possessing coarse-grain information (to the granularity of the GFA) and the GFA possessing more fine-grain information (to the granularity of the RFA). In contrast, with regular Mobile IP, the HA itself possesses the fine-grain information, to the granularity of the FA. The cost of maintaining this information in regular Mobile IP is potentially high handoff latency and high global signaling overhead.

Each Mobile IP registration goes through an RFA and then through a GFA enroute to the HA. When an MH moves between RFAs that are under the same GFA, the GFA does not need to inform the HA that the MH has changed its location. Only when an MH moves between RFAs that are under different GFAs does the GFA need to inform the HA that the MH has changed its location. The HA then behaves as it would in normal Mobile IP when an MH moves between FAs, which means it stops tunneling to the old FA and begins tunneling to the new.

The two-level hierarchy also can be generalized into a hierarchy with three or more levels, in a straightforward way.

6.4.2 Host-Based Routing Schemes

Routing schemes can be host-based or group-based. With host-based routing schemes, as the name implies, forwarding behavior is specified separately for each host. In group-based routing schemes, forwarding behavior is specified for groups of hosts and the same treatment is given to any member of a group (regular IP routing uses such a routing scheme, as explained in Chapter 2). In host-based routing schemes, nodes route packets according to tables or caches indexed by unique host identifiers like their IP address. The most prominent host-based micromobility schemes are cellular IP [12] and HAWAII [13]. The two schemes have many similarities and so we choose to use cellular IP for illustration. Note that despite the name, cellular IP is not directly related to cellular networks. For more details on HAWAII, and comparisons between cellular IP and HAWAII, the reader may refer to [14].

In micromobility zones where cellular IP is used, cellular IP uses forwarding caches containing forwarding information for individual hosts. The information is distributed so that each node knows only how to forward to the next node. The ease and speed with which handoffs are handled (by updating only a subset of the nodes, as appropriate) are an advantage of cellular IP.

6.4.2.1 Protocol Description

The micromobility zone in cellular IP is also called a cellular IP domain. Each cellular IP domain consists of a gateway router and other cellular IP nodes. A cellular IP domain is considered a private network separate from the Internet, having its own way to forward packets, but it can be interfaced with other networks such as the Internet. The gateway router is the interface between the cellular IP domain and external networks like the Internet. An example layout of cellular IP nodes beneath a gateway router is shown in Figure 6.17. The cellular IP nodes are arranged in an inverted tree structure so that each node has only one parent (either another cellular IP node or the gateway), and may itself be a parent node. Conversely, each node is one of the child nodes of its parent node. A node's parent node is one link closer to the gateway than the node itself (and the parent might itself be the gateway) and the path from a child node to the gateway always goes through its parent node. Nodes without children but with wireless interfaces through which they can communicate directly with MHs are known as base stations. The coverage areas of base stations are known as cells. It would be expected that cells in a Cellular IP domain cover a large geographical region, so that micromobility handoffs are more common than macromobility handoffs.

We can think of there being two directions that packets could travel: upstream or downstream, upstream being towards the gateway and downstream being away from the gateway. Cellular IP nodes have upstream interfaces that bring packets closer to the gateway, and downstream interfaces that take them further from the gateway. Since every cellular IP node has one parent, it has exactly one upstream interface and zero or more downstream interfaces. A node can easily discover its upstream interface by receiving beacon messages sent periodically down by the gateway, and recording the interface through which these messages arrive, as the upstream interface. All the other interfaces are the downstream interfaces.

Now we turn to how packets are forwarded in a cellular IP domain. First, we consider the upstream direction. Since each node, including each base station, knows its upstream interface, it is a simple matter for packets from an MH to be forwarded successively through a base station and then through one or more Cellular IP nodes towards the gateway. Each node does not know the full path to the gateway, but only the upstream interface, which is sufficient, since its parent knows how to forward the packet further upstream. When the gateway router gets the packets sent upstream, it then puts them out onto the Internet or back down the cellular IP domain, depending on the destination address.

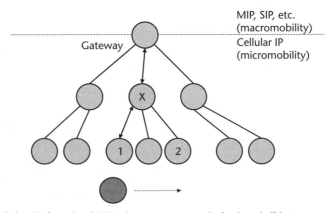

Figure 6.17   Cellular IP domain. "X" is the crossover node for handoff between nodes 1 and 2.

For the downstream direction, each cellular IP node maintains a set of forwarding cache entries. Each entry is an association between an MH (typically, identified by its IP address) and one of the downstream interfaces, and no MH has more than one entry in any given forwarding cache. Thus, when a packet arrives at a cellular IP node from its upstream interface, it searches its forwarding cache for the unique entry corresponding to the destination MH, and forwards the packet to the appropriate interface. Now, how do the cache entries get in the cache in the first place? When an MH first accesses the network through a particular cellular IP base station, it sends a *route update* message to the gateway. This update message gets forwarded upstream from node to node until it arrives at the gateway. Each of these nodes along the path from base station to gateway will obtain the IP address of the MH from the source IP address field of the packet. Nodes can therefore create an entry for that MH, associating the IP address with the interface through which the update arrived. Given the frequent changes in point of attachment to the network, the cache entries have a soft state and need to be periodically refreshed. The cache entries can be updated by regular data (with the source IP address of the MH, and arriving through the expected downstream interface). However, since upstream data is not always being transmitted, a route update would be transmitted by an MH to refresh the cache entries after a period of no upstream transmissions from that MH.

Figure 6.17 shows an MH (the node at the bottom of the figure) moving from the cell covered by base station 1 (BS1) to the cell covered by base station labeled (BS2). The node marked "X" is the crossover node for this handoff; in other words, "X" is the lowest node that is in the upstream of both of the paths from the two base stations to the gateway (so in the "worst case," the crossover node is the gateway). The MH initiates handoff by sending a route update to the gateway through the new base station, BS2. As this route update propagates up to the gateway, for each node below the crossover node, the routing cache may not have any entry for the MH, and the appropriate entry is added. At the crossover node, the critical switch occurs as the cache entry for the MH is updated to point towards the interface heading towards BS2.

6.4.3   802.11 WLAN

Recall from Chapter 3 that 802.11 WLANs can be used as independent base service sets (IBSSs) or extended service sets (ESSs). In an IBSS where every station is in "ad hoc mode" and communicates directly with every other station, the concept of micromobility becomes moot. Either a station is in an IBSS or it is not; there is no mobility tracking issue within an IBSS. In an ESS (which is more common), on the other hand, there are a number of APs all connected to a DS. According to the specifications, the ESS is supposed to look like an ordinary IEEE 802-style WLAN to the higher layers. Thus, even when two MHs are connected to the ESS through two different APs, the APs are supposed to work together such that it looks to the higher layers like there is a direct LAN-like connection between the two MHs. Furthermore, the ESS needs to handle the movement of one or both of the MHs between APs while still maintaining this LAN-like connection.

In the great majority of cases, APs are connected to Ethernet LANs, so the DS is typically an Ethernet LAN. According to the 802.11 specifications, the DS (including the APs) needs to provide five services: association, disassociation, reassociation, distribution, and integration. A station associates with an AP in order to communicate through that AP; compared with Ethernet, the association procedure is analogous in some ways to plugging an Ethernet cable into a hub, because it sets up the basis for communications between the station and that AP. Each station can only be associated with one AP at a time. Disassociation is the procedure that removes the current association. Reassociation is for mobility between APs in an ESS. These three procedures and their related signaling between stations and APs are defined in 802.11. Note that what is specified for reassociation is very minimal, without considerations for how the two APs may cooperate to reduce disruptions in service to the higher layers. As a result, different vendors of 802.11 equipment have implemented proprietary solutions for "smooth handoffs" that do not lose any packets. However, interoperability between APs from different vendors remains a problem. Thus, the 802.11f enhancement to 802.11 attempts to specify certain interactions between APs known as Interaccess Point Protocol (IAPP), especially for handoffs (see Figure 6.18).

The other two functions, distribution and integration, are not defined in 802.11. Integration has to do with wired LANs, which is not an issue when the DS itself uses Ethernet. As for distribution, the Ethernet distributes all frames in the LAN to all the APs. It may appear that the APs could simply act as traditional bridges that connect LAN segments into a LAN, and forward all traffic from one side to the other side and vice versa. However, given that the Ethernet would often have a higher capacity than the link between the APs and the stations, simple bridging implies that the wireless links from each AP would often be congested with traffic not destined to any of the stations supported by that AP. Therefore, a learning table is a necessity—only traffic intended for stations associated with an AP would be forwarded by the AP from the wired side to the wireless side. While this is the simplest solution, other solutions are also possible. These may solve some other problems as well, such

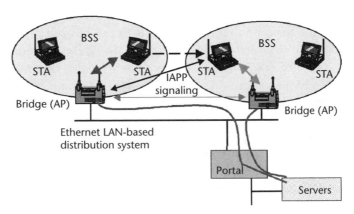

Figure 6.18 WLAN micromobility between APs within an ESS.

as differences in maximum frame size between the wired and wireless sides (an 802.11 frame may be larger than an Ethernet frame, and so it may need to be split into two frames on the wired side [15]).

6.4.4 GPRS

GPRS is used in 3GPP-specified networks. A complex packet-switched core network infrastructure is used, including IP routing internal to the GPRS core network (but "underneath" the user's IP packets), that forms a large L2 mobility-management scheme so that it is only one IP hop between the mobile station and the gateway GPRS support node (GGSN). Interestingly, part of this L2 mobility-management scheme involves tunneling over an IP-based network that is internal to the GPRS network. We will give a fuller description of GPRS and mobility management for GPRS in Chapter 11.

6.4.5 Other Fast Handoff Approaches

While the micromobility approaches are designed to reduce handoff latency by reducing the time it takes for updates and registrations to occur, recent fast handoff work by the IETF attempts to reduce handoff latency by beginning the update and registration process earlier, and to reduce the effect of handoff latency by forwarding packets between the old and new FAs while waiting for the update and registration to complete. These schemes exploit cross-layer triggering (between L2 and L3). In fact, there are two such schemes, one being the preregistration handoff technique and the other the postregistration handoff technique.

In the preregistration handoff technique, the main idea is that the MH starts the registration process (part of L3 handoff) before the L2 handoff is complete. There are a number of possible scenarios, one of which is that an FA communicates with the FAs around it, so it can do a *proxy agent advertisement* on behalf of a potential new FA. This proxy advertisement is basically an agent advertisement except that it is sent on behalf of another FA. It can be triggered by L2 triggers. For instance at the old FA when the signal strength for the MH-FA link drops below some threshold, an L2 trigger may result, indicating an impending L2 handoff to the subnet served by the new FA. Thus, the MH can begin Mobile IP registration with a new FA through its old FA, rather than first waiting for the L2 handoff to complete, and thus reducing handoff latency.

We note that cross-layer triggering breaks one of the original design principles of the Mobile IP design—that it makes no assumptions on L2, in order not to constrain Mobile IP to be usable only where certain assumptions on L2 apply. However, by not making assumptions on L2, Mobile IP cannot take advantage of cross-layer triggering; in particular, the L3 handoff cannot occur until at least the L2 handoff completes and L3 somehow discovers that a L2 handoff has taken place (e.g., by hearing an agent advertisement from a new FA). Furthermore, without incorporating the involvement of both old FA and new FA in the handoff, the original Mobile IP requires registration with the new FA to be completed before the MH can resume communications.

In the recent fast handoff work, the approach to cross-layer triggering is that if and only if cross-layer triggering is available for a given system, it can be used to speed up Mobile IP handoffs by starting the L3 handoff before the L2 handoff has completed. Thus, if a particular network does not support cross-layer triggering, it can still use Mobile IP, even though it cannot support this fast handoff mechanism. Thus, backward compatibility is assured.

6.4.6 Reducing the Impact of Handoff Latency

Many enhancements and modifications of Mobile IP have been proposed, to reduce the amount of handoff latency. Instead of asking how we may reduce the handoff latency, another angle that has been tried is to invert the question. Given a certain amount of handoff latency, what can we do to reduce the number of dropped packets? After all, the fundamental problem with the handoff latency is dropped packets.

The main answer to the turned-around question is to have a scheme to forward packets from the old network to the new network, rather than simply dropping and losing them. Such an idea was first proposed early in the development of Mobile IP—before micromobility and fast handoff schemes appeared on the scene. It was bundled with route optimization, one of the early enhancements to Mobile IP that has already been introduced in Section 6.1.2.4 [4]. The way the forwarding works is that a new extension, the previous foreign agent notification extension (PFANE), is added to the Mobile IP Registration message. The PFANE tells the new FA the address of the old FA and that the MH wishes the new FA to ask the old FA to forward any packets it may have buffered for the MH.

One issue with such packet-forwarding schemes is that even though they reduce the fundamental problem of dropped packets, they may introduce a secondary problem. This is the problem of additional *end-to-end latency,* or additional time for packets to get between the CH and the MH. While this may be insignificant when the old and new foreign networks are close to each other, it may be quite significant in some cases when the packets have to take a long and circuitous route between the two networks (even when the two networks are geographically close, the network topology may be such that a long and circuitous route is unavoidable). Moreover, even if the old and new foreign networks are close to each other, there can be delays in packet buffering, and possibly lengthy delays (proportional to handoff latency) while waiting at the old network for the new network information.

6.5 Summary

This is the first of three chapters covering the big three network-level challenges in wireless IP networks, namely mobility, QoS, and security. This chapter introduces the problems associated with mobility, as well as various mobility concepts, such as handoffs and location management, types of mobility (personal, terminal, session, and service), and L2 versus L3 mobility. The text explains our focus on the subproblems having to do with finding a wireless device (e.g., for routing, session initiation) even though it may move from time to time, whether before or even during sessions. The Mobile IP protocol is introduced as a well-established solution, and assessed. Enhancements such as route optimization, micromobility, and fast handoffs are introduced as ways to correct shortcomings of Mobile IP. Micromobility schemes discussed include hierarchical Mobile IP, host-based schemes like Cellular IP, and intrawireless LAN mobility. We discussed SIP-based mobility as an application-layer alternative to Mobile IP, and TCP migration and MSOCKS as transport-layer alternatives.

References

[1] Vijayan, R., and J. Holtzman, "A Model for Analyzing Handoff Algorithms," *IEEE Transactions on Vehicular Technology*, Vol. 42, No. 3, August 1993, pp. 351–356.

[2] Wong, K. D., and D. Cox, "A Pattern Recognition System for Handoff Algorithms," *IEEE Journal on Selected Areas in Communications*, July 2000, pp. 1301–1312.

[3] Perkins, C., *Mobile IP: Design Principles and Practices*, Reading, MA: Addison-Wesley, 1996.

[4] Perkins, C., and D. Johnson, "Route Optimization in Mobile IP," draft-ietf-mobileip-optim-11.txt, work in progress, 2001.

[5] Perlman, R., *Interconnections: Bridges and Routers*, Reading, MA: Addison-Wesley, 1992.

[6] Schulzrinne, H., and E. Wedlund, "Application-Layer Mobility Using SIP," *Mobile Computing and Communications Review (MCCR)*, Vol. 4, No. 3, July 2000, pp. 47–57.

[7] Snoeren, A., and H. Balakrishnan, "End-to-End Approach to Host Mobility," *Mobicom 2000*, 2000.

[8] Maltz, D., and P. Bhagwat, "MSOCKS: An Architecture for Transport Layer Mobility," *Infocom 1998*, 1998.

[9] Salzer, J., D. Reed, and D. Clark, "End-to-End Arguments in System Design," *ACM TOCS*, November 1984, pp. 277–288.

[10] Wellington, B., "Secure Domain Name System (DNS) Dynamic Update," RFC 3007, November 2000.

[11] Gustafson, E., A. Johnson, and C. Perkins, "Mobile IP with Regional Registration," draft-ietf-mobileip-reg-tunnel-07.txt, work in progress, 2002.

[12] Campbell, A., et al., "Comparison of IP Micro-Mobility Protocols," *IEEE Wireless Communications Magazine*, February 2002.

[13] Ramjee, R., "IP Micro-Mobility Support Through HAWAII," work in progress, July 2001.

[14] Reinbold, P., and O. Bonaventure, "IP Micro-Mobility Protocols," *IEEE Communications Surveys and Tutorials*, Third Quarter 2003.

[15] El-Hoiydi, A., "Implementation Options for the Distribution System in the 802.11 Wireless LAN Infrastructure Network," *ICC 2000*, June 2000, pp. 164–169.

CHAPTER 7

QoS

7.1 Introduction

Computer networking relies on sharing of limited network resources for communications. If communications resources were unlimited, a direct point-to-point link of unlimited bandwidth would be possible between any two end-points that wished to communicate. Clearly, this is not possible in the real world. Therefore, in the real world, two end-points that wish to communicate, unless they are connected to the same LAN, need to communicate through a network of links that have the following characteristics: (1) the links are shared between traffic from multiple sources to multiple destinations, (2) the links have limited bandwidth, (3) the links have propagation latency (i.e., time taken for data to traverse the links), and (4) the nodes directly connected to each link have limited processing power and memory capacity.

How should these limited resources be shared? The easiest way is to treat all traffic the same, meaning all packets are treated the same, accordingly to loose general principles of fairness. Thus, all packets are processed according to the order in which they arrive, in a *first in first out* (FIFO) manner. If so many packets arrive at a node that its buffers are filled, subsequently arriving packets may be dropped. The Internet, like many other networks, behaves like this.

However, this simple treatment of packets is not sufficient for today's *multi-service* networks, which support a variety of services and applications. The different applications have different requirements on the communications services provided by the network. For example, voice and video have very stringent delay and latency requirements, as already discussed in Chapter 4. File transfer, on the other hand, does not have the same delay and latency requirements. These kinds of requirements, plus others such as "guaranteed" bandwidth requirements, are known as *QoS requirements,* and the whole field that is about these requirements and the ways they are met in networks, is known as *QoS.* In the context of traffic over IP networks, Table 7.1 shows some different categories of traffic and their QoS requirements.

Furthermore, if some customers are willing to pay for premium service, the network should be able to give priority to the traffic from these customers, or guarantee certain amounts of bandwidth (e.g., a certain fraction of the bandwidth of the links). This concept can be generalized to a requirement to support multiple classes of service. The requirement for different classes of service need not be related only to

Table 7.1 Internet Traffic Characteristics

IntServ Category [1]	IntServ Subcategory	3GPP Category [2]	Examples	Constant Rate Necessary	Low Delay Necessary	Low Jitter	Low Delay Preferred
Tolerant Real-Time		Streaming	Audio/video streaming	√	√		
Intolerant Real-Time		Conversational	IP telephony, teleconferencing	√	√	√	
Elastic (Non-Real-Time)	Interactive Burst	Interactive	Telnet, HTTP				√
	Interactive Bulk Transfer		FTP				√
	Asynchronous Bulk Transfer	Background	SMTP				

the paying capacity of customers. In an enterprise setting, for example, it could also make sense to prioritize various traffic based on business needs and requirements. For example, urgent business videoconferencing traffic could be more important than casual Web browsing in the same enterprise network. In a military context, a general's communications could be given priority over lower-ranking officers' communications.

A variety of QoS mechanisms can be, and are, implemented in various networks to support QoS. We will illustrate in the context of IP networks in Section 7.2.

7.2   IP QoS: Mechanisms

We now examine schemes for providing QoS in IP networks. Later, in Section 7.4, we will particularize to wireless IP networks.

7.2.1   Introduction to IP QoS

The existing Internet is largely a best-effort network. No guarantees are provided that packets will not be lost, that packet propagation times will be constrained, that packets will travel along a particular path, or that packets will arrive in sequence. Therefore, it is said that the Internet provides *best-effort service,* or that it does its best to move the packets from source to destination. As the Internet has evolved from a research network used primarily by the scientific and research community to include significant levels of usage by the general public as well as corporations, many in the business community are demanding better service quality and have indicated a willingness to pay for it. Furthermore, as the range of applications using the Internet increases, there is an increasing difference in the network-level requirements of these applications. Some general classes of such application traffic requirements are shown in Table 7.1, from which it can be seen that different network-level treatment may be required for different types of traffic. Different QoS may be required for different types of traffic.

The telecommunications industry is developing mechanisms to support service differentiation by assigning different levels and types of service guarantees to different traffic flows, including guaranteed quicker delivery time or higher probability of availability of desired throughput. The service differentiation requirement is the fundamental requirement for IP QoS. Because of high network utilization, high levels of service quality cannot be provided to every network user. The IP QoS solution is predicated on three fundamental requirements:

1. To provide mechanisms to offer different levels of services to different users, even though some users might get better service than others. Thus, IP QoS is inherently "unfair."
2. The service should not usually be completely denied to lower-priority traffic.
3. The mechanisms are efficient, not wasteful of valuable network resources.

Therefore, the current trend on IP QoS is towards controlled and efficient unfairness.

7.2.2 Scope and Outline

In this chapter, our focus is on the basics of providing QoS, and IP QoS in particular, and we also consider wireless IP QoS. QoS-related topics that are excluded from the scope of this chapter include measurement (whether by end-system users or network operators), pricing and billing models, and security-related aspects.

Here, we briefly describe some of the common QoS mechanisms in use today. These mechanisms can be found in routers and other network equipment. However, it is essential that the QoS mechanisms be coordinated into a coherent architecture for the provision of QoS. That is the role of QoS frameworks like integrated services (IntServ) and differentiated services (DiffServ), as discussed in Section 7.3.

QoS mechanisms include various protocols and procedures that can be used as building blocks for a network architecture that provides QoS to handle packet transfers. The descriptions that follow are brief and at a functional, conceptual level. A description of *implementation* of token buckets, for example, is out of the scope of this document.

7.2.3 Requirements

Some of the main requirements that drive the need for various QoS mechanisms have to address the following:

- How to reserve network resources for certain flows (Section 7.2.4);
- How to control the admission of traffic into a given network (Section 7.2.5);
- How to control the sequence packets processed by a node, or sent out by a node, in other words, input or output queuing (Section 7.2.7);

- How to control the rate of flow of traffic, whether on an aggregated, flow-by-flow, or class-by-class basis (Section 7.2.8);
- How to monitor compliance of flows to profiles (Section 7.2.9);
- How to control the routing of different flows in the network (Section 7.2.10).

In these networks, decisions need to be made on whether the resource reservation requests can be satisfied. This is a form of *admission control*. Admission control is used, in combination with other QoS mechanisms, to control traffic load in a network. These other mechanisms include flow rate control mechanisms, routing control/traffic engineering, and others. Admission control ensures that the amount of traffic allowed to use the network is controlled. Once traffic is admitted to the network, rate control mechanisms, as well as other mechanisms, help to prevent congestion.

7.2.4   Resource Reservation

Perhaps a natural first step in moving away from a purely best-effort-service Internet is to provide a way to *reserve resources* for selected users while still using traditional IP routing. Resource reservation refers to dynamic arrangements for QoS provisioning[1] that can be made end-to-end over one or more intermediate networks. Usually, a reservation is specified by a *traffic profile*, a description of the rate, burst size, and other features of the reserved traffic. The traffic profile may be used by some of the flow rate control mechanisms described in Section 7.2.8.

7.2.4.1   Resource Reservation Protocol (RSVP)

RSVP is a signaling protocol that allows resource reservation to be coordinated in a network [3]. RSVP signaling generally results in resources being reserved in RSVP-capable routers in the transit path of a traffic flow. Each of these routers maintains a soft state that needs to be periodically refreshed. RSVP is unidirectional—resources are reserved in one direction from source to destination only. Traditionally, RSVP has been very closely tied to the IntServ QoS framework (see Section 7.3).

Resources are reserved using the following procedures (see Figure 7.1). The source node sends *Path* messages towards the destination node. This allows a soft-state path characterization to be set up in each of the RSVP-capable transit nodes between the two end-points. (In general, whether for RSVP or not, soft state is state that must be refreshed, whereas hard state does not need to be refreshed.) However, resources are not actually reserved at this time; they are not even requested at this time. Instead, it is only when the destination node sends back *Resv* (reservation request) messages toward the source (in the same path but in the reverse direction) that the QoS services are actually requested. This is because RSVP is designed for the receiver to be responsible for requesting QoS services rather than the sender. So, the

---

1.   Note that from a broader perspective, the philosophy and time scales of resource allocation in the network are major differences between the IntServ and DiffServ frameworks (as will be discussed later).

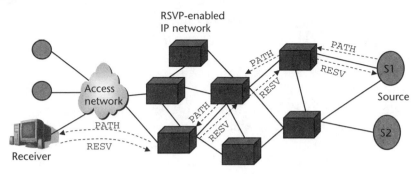

Figure 7.1   RSVP.

forward Path messages are meant to set up in each intermediate router the soft state information about the path back to the sender and to inform the receiver what kind of traffic is to be sent so it can make the necessary reservation requests. Note that the reservation requests can be shared or distinct (i.e., shared between traffic from more than one sender, or just for traffic from one sender). This is one reason why RSVP assigns the receiver responsibility for making the reservation requests. The receiver can figure out best what resources are needed to support the possible combination of traffic from multiple senders, whereas each sender only knows about the traffic it is sending.

A problem with RSVP is that RSVP messages are special transport layer packets, like Internet Control Message Protocol (ICMP) packets, that are processor intensive for core routers. This makes it impractical to support every QoS-enabled session with RSVP signaling.

7.2.5   Admission Control

Admission control refers to the decision whether to admit a certain flow or aggregation of flows into a network, in response to a request for resource reservation. It makes sense for admission control decisions to be made by an entity with an overall view of network resources. This is the idea behind bandwidth brokers, to be discussed in Section 7.2.5.1.

We note that mechanisms like token buckets and shapers placed at the edges of the network can be thought of as a form of *soft* admission control since they may just slow down the rate of traffic entering a network, rather than completely deny access. However, for the purposes of this chapter, we classify such mechanisms under traffic shaping in Section 7.2.8. The reason is the operational similarity between such mechanisms and more traditional rate control mechanisms like queues.

7.2.5.1   Bandwidth Brokers

There are two basic choices for dynamic allocation of bandwidth resources in the routers in a given network. The first is that each router makes its own decisions, perhaps as configured manually or by policy. The second is that a single entity makes the decisions for the network, taking into account the dynamic conditions in the network. This entity is often called a bandwidth broker.

By the nature of its functions, it is typical that each domain would have its own bandwidth broker to control its bandwidth resources. It would be involved in negotiations (perhaps with another domain's bandwidth broker) for resource allocation for interdomain traffic, making decisions on, for example, whether or not to admit resource reservation requests. A functional description of bandwidth brokers in the context of a DiffServ framework has been proposed for interdomain traffic by Nichols et al. [4]. Figure 7.2 shows an example of a bandwidth broker providing admission control services to an administrative domain. In this example, we show common open policy service (COPS) as the communication protocol between the bandwidth broker and the edge routers of the network, but other protocols could also be used, including Simple Network Management Protocol (SNMP). COPS is "a simple client/server model for supporting policy control over QoS signaling protocols" [5]. It is also a simple query and response protocol for use between a policy server (such as the bandwidth broker in this case), and a client (the edge router in this case). Unlike most SNMP implementations that use UDP as their transport protocol, COPS uses TCP for reliable message exchanges. Note also in Figure 7.2 how the routers internal to the domain are not involved in the admission control process (and rightly so, as admission control should be done at the entry points to the domain).

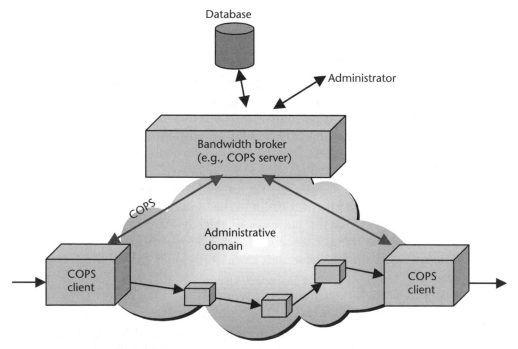

Figure 7.2 A typical bandwidth broker architecture.

7.2.6   Packet Classification and Marking

Packet classification is the separation of packets into different classes based on some criteria. For example, the classification may be port-based (the input and/or output port of the packet), transport protocol–based (e.g., TCP or UDP), application-based (such as HTTP-based), or address-based (based on the source or destination address of the packet). It is also possible that out-of-profile packets are reclassified, typically into a lower priority class. Marking is the setting of bits in the packet header corresponding to the packet class. The bits marked could be the three bits in the Type of Service (ToS) field and the three bits in the precedence field of the IP header.

Packet classification and marking allows differential treatment of different packet classes using other, complementary QoS. Packet classification and marking may be handled by external sources like a customer or other network. When the packets enter a new network, that network can accept the classification or reclassify. The decision depends on the QoS architecture and policies used.

7.2.7   Queuing Disciplines

Queuing is fundamental to IP QoS schemes because each router has one or more input queues and one or more output queues that may be where the bulk of "processing time" is spent. The use of various queuing disciplines can be used to change such functions as processing times and flow rates. Most of the queuing disciplines discussed here operate on router output queues, although priority queuing could be implemented to operate jointly on input/output queues.

7.2.7.1   FIFO Queuing

FIFO queuing is the standard, basic queuing discipline. All packets are treated equally without preference except that earlier arriving packets leave before later arriving packets. Advantages of FIFO queuing include:

1. Highly optimized forwarding performance resulting from years of experience by router manufacturers.
2. For lightly loaded networks with sufficient transmission and switching capacity, queuing is necessary only for smoothing intermittent traffic bursts. FIFO does this very efficiently.

Disadvantages of FIFO queuing include:

1. For networks that are not lightly loaded, FIFO queuing may result in loss of packets through discarding when the queues are full.
2. FIFO does not allow certain packets to receive priority/preference over other packets.

Hence, for lightly loaded networks, FIFO is efficient for smoothing traffic bursts. However, in cases where the network is more heavily loaded (cases needing

QoS mechanisms), there should be a way to differentially set the speed of degradation of various flows when queues start filling up and packets get dropped.

7.2.7.2   Priority Queuing (PQ)

PQ (see Figure 7.3) is a non-FIFO queuing discipline. The router examines the input queue, takes high-priority packets and places them in the output queue ahead of normal packets. This effectively reorders packets according to priority. Advantages of PQ include:

1. The number of priority levels can be set flexibly based on need.
2. The granularity of each class is flexible.

Disadvantages of PQ include:

1. Computational overhead and impact on packet forwarding performance, especially as the number of priority levels increases.
2. Buffer starvation—lower-priority traffic may be queued for very long periods, and/or be mostly dropped, when higher-priority traffic volume is high. Thus, low-priority traffic is effectively denied service. It may be better to provide reduced levels of service to the low-priority traffic instead, as is the case for the following queuing schemes.

7.2.7.3   Fair Queuing (FQ), Weighted Fair Queuing (WFQ), and Class-Based Queuing (CBQ)

Larger traffic flows present more packets to queue inputs. With FIFO queuing, these flows also comprise a proportionately larger portion of the outputs. Is it unfair for larger traffic flows to "starve" smaller traffic flows in this way? FQ is based on the notion that it is. It attempts to balance out traffic-flow volume at the queue output regardless of flow volume at the input, by using per-flow queues and interleaving between the queues. Therefore, FQ results in preferential treatment to low-volume traffic flows. WFQ is a variation and generalization of FQ that weights the outputs

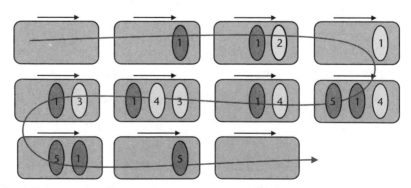

Figure 7.3   Priority queuing illustrated.

of the per-flow queues (according to the IP ToS field, for example) as well. Variations of WFQ include worst-case WFQ, start-time FQ, self-clocked FQ, and hierarchical WFQ. CBQ is another member of the FQ family (although attempts have been made to differentiate it from WFQ, there is no clear distinction between WFQ and CBQ that everybody agrees upon).

Advantages of WFQ include:

1. It prevents a misbehaving TCP session from consuming a large fraction of resources at the expense of other flows.

2. The fairness aspect of this type of scheme prevents buffer starvation.

Disadvantages of WFQ include:

1. WFQ is a way to approximate generalized processor sharing (GPS), where the link sharing between queues is accomplished by scheduling alone.

2. It does not scale well. Computational overhead is high, and forwarding performance may be affected.

The various queuing disciplines are compared in Table 7.2.

7.2.8   Traffic Shaping

Traffic shaping refers to controlling the rate of traffic passing through a given router. It is often used at the ingress points of a network as a form of soft admission control. Leaky bucket traffic shaping constrains the rate to be less than a maximum value, while token bucket traffic shaping constrains the *average* rate to be less than a maximum value.

7.2.8.1   Leaky Bucket

In a typical FIFO queue, as soon as a packet leaves the queue, the next one (if any) is ready to leave. A leaky bucket (see Figure 7.4) can be thought of as a generalized FIFO queue by simply adding a fixed time interval between the output of packets from the queue. It can be pictured as a bucket with holes at the bottom, through

Table 7.2   Comparison of Queuing Disciplines

	FIFO Queuing	Priority Queuing	WFQ/CBQ
Allows flow differentiation	No	Yes	Yes
Avoids buffer starvation	Yes	No	Yes
Favors low-volume flows	No	Not necessarily	Yes
Link bandwidth division	No	No	Yes, good performance

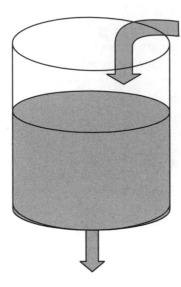

Figure 7.4 Leaky bucket.

which water leaks at a fixed rate depending on the size of the holes.[2] Advantages include:

1. It is simple to implement, almost like a FIFO queue, and can take advantage of the design experience router manufacturers have in FIFO queuing.
2. It works well where FIFO queuing works well, and has a controllable output rate, whereas FIFO queuing is like a leaky bucket always fixed at maximum output rate.

Disadvantages include:

1. A leaky bucket is good for restraining traffic leaving a router. However, existing leaky buckets use a fixed leak rate [6]. Hence, it cannot adapt to network conditions allowing a higher leak rate.
2. It is not as useful in cases where the average rate is of interest, rather than the maximum rate.

7.2.8.2 Token Bucket

A token bucket is a more flexible traffic shaping mechanism than a leaky bucket, because it allows traffic bursts over the nominal desired long-term average traffic rate. It can be pictured (see Figure 7.5) as a bucket into which tokens are added from a fixed rate source. Whenever a packet from the associated flow is at the head of the

---

2. The analogy is not perfect: Unlike water leaking from a real bucket, where the rate depends not just on the size of the hole but also on the depth of water in the bucket, the output rate of leaky buckets for traffic shaping generally does not depend on the queue size.

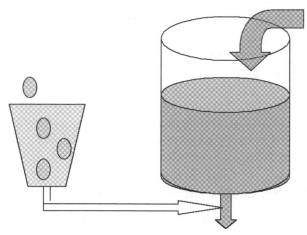

Figure 7.5   Token bucket.

queue and ready to be transmitted, it is transmitted only if there are enough tokens for it. These tokens are then removed from the bucket. If there are not enough tokens to support a transmission, the queue waits for more tokens to be added until there are enough. Clearly, the token bucket allows large bursts of traffic to flow through as fast as the queue will output them, after a period of reduced activity in which a large number of tokens have accumulated. This is a form of memory and it allows bursts of traffic to exceed the nominal desired long-term average rate. Key parameters are token rate, token bucket depth, peak traffic rate (packets per second), and the range of expected traffic packet sizes. Advantages of using token buckets include:

1. It does not artificially restrict flow rates to predetermined levels.
2. It allows for more flexibility in shaping variable-rate traffic (is less strict, as long as the average rate is acceptable).

Disadvantages include that it is not as useful in cases where the maximum rate is of interest, rather than the average rate.

7.2.9   Policing

There are multiple definitions of policing in the literature, which all have to do with treatment of packets that are nonconforming (with a profile). According to one definition, policing refers to the process of dropping packets from a flow that are nonconforming [7]. This prevents the flow from adversely affecting other traffic through the node. Another definition includes the possibility that the nonconforming packets are not necessarily always dropped, but marked as nonconforming and treated differently from conforming packets, for instance, being treated with best-effort rather than with any guarantee of service. A broader definition of policing considers the previous two definitions to be somewhat narrow. For example, those definitions might be considered "simple policing," whereas policing in general can include traffic shaping (which, if successful, does not mean that the nonconforming packets are lost or marked for poorer treatment) [8].

The kind of token bucket described in Section 7.2.8.2 is sometimes referred to as a *shaping token bucket,* in contrast to a *policing token bucket* [4]. The difference is what happens when there are no tokens or not enough tokens. In a shaping token bucket, the packet waits till there are enough tokens for it to pass through. In a policing token bucket, the packet is immediately dropped, or otherwise policed (e.g., reclassified and remarked).

7.2.10 Routing Control and Traffic Engineering

In the existing Internet, different packets of a flow may be routed along different paths, and packets from different flows are treated equally in routing and forwarding decisions, regardless of the QoS requirements. One of the ways to provide QoS guarantees like bounded delay is to specify the routing and forwarding of packets of specific flows to meet the QoS criteria. Controlling the routing of traffic in this way is known as traffic engineering. The basic functional requirements for traffic engineering are [9]:

- Distribution of topology information, to allow nodes to build correct topology maps and to assist in path selection;
- Path selection, to select a path[3] based on some criteria, such as bandwidth, delay, shortest path;
- Directing traffic along computed paths, using forwarding tables (computed independently at each node using traditional IP routing protocols or signaled protocols such as MPLS).

MPLS is a solution for the third requirement, whereas QoS routing (also known as constraint-based routing) is based on the first two. MPLS is out of the scope of this book, but we briefly describe QoS routing here.

Most research in QoS routing is based on finding a path for a connection between two end points. The path should satisfy a set of constraints, such as link or path constraints, and can be optimized over links or paths for a certain resource (e.g., maximum bandwidth). The constraints may be bandwidth constraints or delay constraints, for example (or energy constraints for wireless IP networks). One problem with QoS routing mechanisms is state maintenance, which is the problem of what local or global state information should be maintained at each node. The major problem, though, is that of finding routes through the network, in the presence of constraints and desired optimizations. Some of the problems are of polynomial complexity, but some are NP-complete.

---

3.   In the case of multicasting, this would be a tree, rather than a path.

7.3   IP QoS Frameworks

The mechanisms discussed in Section 7.2 are often used in combination in networks, in order to provide end-to-end QoS, or to enforce some QoS policies in a network domain. Routers typically combine some of these mechanisms, for example as shown in Figures 7.6 and 7.7.

There should be ways to more systematically coordinate the various QoS mechanisms used in a network according to some kind of framework to allow end-to-end provision of QoS for a variety of network traffic, consistent with some idea of network QoS policy. IntServ and DiffServ are two such frameworks.

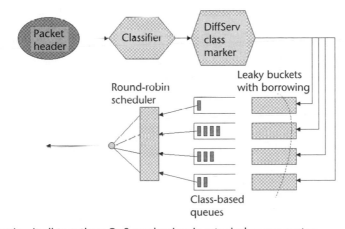

Figure 7.6   Putting it all together; QoS mechanism in a typical access router.

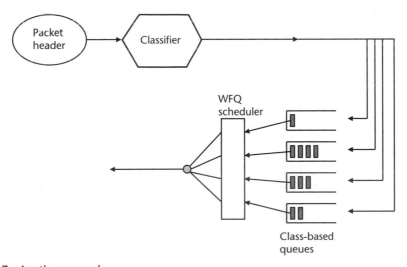

Figure 7.7   Another example.

7.3.1 IntServ

The IntServ framework for QoS was published in 1994 [1], where the name comes from the fact that it attempts to provide service for different types of nonreal-time and real-time traffic on the same network. It uses admission control, rate control mechanisms, and resource reservation mechanisms to deliver QoS. The IntServ model takes into consideration the following issues:

- QoS requirements for different types of traffic, that can be met by service differentiation;
- Resource reservation.

In addition to the main features of IntServ, several other features have been proposed. In cases of link sharing (e.g., between traffic of different protocols), WFQ is proposed. Packet dropping is also proposed, for example by allowing low-priority packets to be preempted. Provision for usage feedback (also known as accounting) is proposed but left up to the implementation, as it is considered a contentious issue.

7.3.1.1   Service Classes

The original IntServ framework [1] did not specify in detail the kinds of service classes that could be supported by IntServ. More recently, there have been attempts made to address this deficiency. RFCs have been written on supporting various service classes using the IntServ model, for instance, the controlled load service class and the guaranteed QoS service class [8, 10]. Each class has its own QoS requirements.

The controlled load service class is recommended for jitter-tolerant real-time applications. The QoS requirement for such traffic is that the network behaves like a lightly loaded network that uses best-effort delivery, whether or not the network is in fact lightly loaded. The network behaves like a lightly loaded network at all times in the sense that the probability of successful delivery is very high and the transit delay experienced by most packets has little jitter. The way it does this is through admission control. To send traffic through such a network, reservation requests must be made, possibly by RSVP. Each network element accepting a request for controlled load service must ensure that it has sufficient bandwidth and packet-processing resources to handle the requested traffic profile (specified in terms of TOKEN_BUCKET_TSPEC, which is the collection of key parameters for token buckets in Section 7.2.8.2). No numerically precise definitions are given for what sufficient means in this case, except for guidelines on what kind of best-effort service a lightly loaded network can provide. If the actual traffic in a flow exceeds the requested profile, the network element can drop packets or use other means to ensure that other flows are not adversely affected. For example, WFQ may be used, where the weights are based on the reservations. Also, the RFC specifies how to police or shape each flow based upon their requested profiles.

The guaranteed QoS service class is for applications with a bandwidth guarantee and a delay bound that are jitter-intolerant. End-to-end delay depends both on queuing delay and transmission delay. The guaranteed QoS service class aims to control the maximum queuing delay (however, no guarantees are made on transmission delay). Basically, the concept is that given the TOKEN_BUCKET_TSPEC requested by a flow, a network element can compute the maximum queuing delay. Hence, guaranteed service can guarantee that if traffic for a specific flow stays in profile, the maximum queuing time is fixed, and packets will not be discarded due to queue overflows. Although the maximum queuing delay is guaranteed, many packets will arrive earlier, and the jitter-intolerant application needs to buffer the early arriving packets until it is ready to process them. For out-of-profile packets, policing and reshaping are two alternative treatments. It is recommended that policing be performed at network edges, and reshaping in the rest of the network. The RFC specifies how to use a token bucket to reshape the traffic.

7.3.2 DiffServ

The IntServ framework has a major problem in that it is not scalable as the size of a network grows. Resource-reservation signaling is performed for every flow. The resource reservations need to be refreshed periodically. Furthermore, core routers in a network must keep track of the state of every flow that passes through them and process the RSVP messages. Figure 7.8 shows the IntServ model when there are just two flows between PCs (on the left) and servers (on the right). Each flow is treated separately. It is easy to imagine how we could have thousands of such flow lines if there are thousands of sessions; this shows that IntServ does not scale well for large networks with many flows.

The DiffServ (or DS) framework is an attempt to address the shortcomings of IntServ. The main idea is that it is not necessary to make separate resource reservations for every flow. Instead, there can be a few service classes, and each flow can belong to one of the classes. This allows grouping of traffic passing through a network element into aggregates, each of which belongs to one class. All traffic in an aggregate (rather than in each flow) is handled the same way. Furthermore, the treatment of each service class can be preconfigured in the core routers, based on the expected levels of traffic of each class, reducing the dynamic changes that occur even in core routers with IntServ. Additional processing of the packets, such as policing

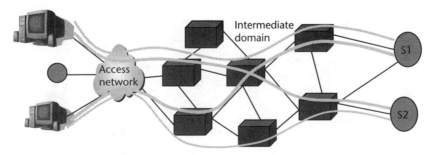

Figure 7.8 IntServ, with per-flow reservations and treatment, lacks scalability.

and shaping, can be performed at the edges, and packets marked according to the appropriate class of service. The DiffServ philosophy is to keep the forwarding path simple, push complexity to the network edges as much as possible, and provide a service that does not make assumptions about the type of traffic using it, while allowing the dominant Internet traffic model to remain best-effort. Figure 7.9 shows an example in which all traffic is aggregated into two classes in the core network where DiffServ is used. Clearly, adding many more flows to the network edges will not significantly increase the complexity in the core network since the same two classes would still be used.

The IETF DiffServ working group has provided an architectural framework for DiffServ. It has redefined the IPv4 header ToS octet as the *DS field* (in the case of IPv6, it is the traffic class octet that maps to the DS field). Packets are marked in the *differentiated services code point* (DSCP) field of the DiffServ field, the rest of the DiffServ field being currently unused. The DSCP field is currently six bits. Traffic aggregates are flows whose packets all have the same DSCP field *behavior aggregates*. Most of the computationally intensive functions including classifying, marking, and shaping (in general, *traffic conditioning* with a variety of QoS mechanisms) are performed at edge routers of a *DS domain*. In the interior routers of a DS domain, packets only have to be classified by their DSCP field into behavior aggregates. The behavior aggregates are treated according to the *per-hop behavior* (PHB) associated with their DSCP markings. The PHB is the externally observable forwarding behavior applied at a DiffServ-compliant router to a DiffServ behavior aggregate. Together with the processing at the edge routers, the PHBs are used to provide different *services,* where a service is the overall treatment of a defined subset of a customer's traffic within a DiffServ domain or end-to-end.

PHBs that have been proposed include assured forwarding (AF) and expedited forwarding (EF) [11, 12]. AF is a means for a service provider DS domain to offer different levels of forwarding assurances for IP packets received from a customer DS domain. AF is actually a 12-member PHB group, divided into 4 classes and 3 levels of drop precedence within each class (more recently, it has been proposed to say that AF is a *type* of PHB group with 3 members, and each of the 4 classes is an *instance* of the AF type [13]). Each AF class is allocated some forwarding resources (bandwidth and buffer space) in each DiffServ node. Each class should be configured to receive

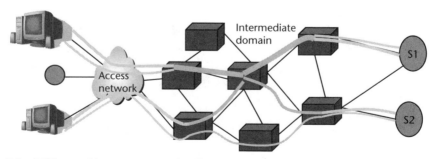

Figure 7.9 DiffServ, with aggregation of traffic (into two classes, in this example).

the desired bandwidth over both large and small time scales. Meanwhile, EF configures nodes so that a behavior aggregate has a well-defined minimum departure rate (well defined in the sense of being independent of the intensity of other traffic at the node). Together with appropriate conditioning of the behavior aggregates at the boundary routers, the so-called premium service may be provided, using EF.

7.4   QoS in Wireless Networks

In wireless networks, there are a number of issues related to providing QoS. First, there are the issues associated with all wireless networks regardless of host mobility, because of the characteristics of the wireless link, including relatively lower bandwidth, higher latency, and higher errors than comparable wired links. Second, there are the additional complications that come with mobility. We illustrate some of the issues by exploring QoS support in wireless LANs in Section 7.4.1, and then consider the impact of mobility on QoS support in Section 7.4.2.

7.4.1   WLAN QoS Support

In Chapter 3, we introduced WLANs, and in particular the IEEE 802.11 WLANs, popularly known as wi-fi. We mentioned the hidden terminal and exposed terminal problems in Chapter 3, and that the CSMA/CA medium access scheme is used in 802.11 to handle medium access while taking care of these challenges. We are now ready to discuss the 802.11 MAC in more detail, including the CSMA/CA scheme. This is directly relevant to 802.11 QoS, as the 802.11e addendum to 802.11 enhances the original 802.11 MAC to support QoS.

The 802.11 MAC consists of a distributed coordination function (DCF) and a point coordination function (PCF). The DCF is implemented in all stations and can be used in both ad hoc and infrastructure modes. The PCF, on the other hand, only makes sense when there is a point coordinator, namely an access point, and hence it only is used in infrastructure mode to coordinate the traffic in each BSS within an ESS. When there is an AP involved, the MAC alternates between contention periods (CPs) and contention free periods (CFPs), where DCF is used in CPs, and PCF in CFPs. However, DCF and PCF share the same underlying MAC structure and mechanisms, with just a few additions for PCF. The same CSMA/CA mechanisms (physical and virtual carrier sensing, random back-off after busy medium detection, differential interframe spacing, and positive ACKs) are used in both PCF and DCF, although the PCF additionally builds upon these mechanisms to seize control of the medium during CFPs. Since the PCF is built upon the fundamental services provided by the DCF, we describe the DCF first.

7.4.1.1   DCF

In CPs, where DCF is used, a station that does not have a packet to send will still be continually monitoring the wireless channel, to perform carrier sensing. These carrier-sensing mechanisms, as we will see, may allow the station to predict how long the medium will be busy. Thus, when the station has a packet to send, there are two cases: the determination that the medium is not busy and the determination that the medium is busy. In the first case, it waits for a time interval DIFS (DIFS are part of the interframe spacing scheme that we will consider in Section 7.4.1.2, since it is only fully utilized with PCF) and then transmits. In the second case, however, it first waits till the medium is not busy, waits for DIFS, and then waits an additional random back-off time (to avoid clashing with other stations that might have been silently waiting at the same time). The random time is drawn from a uniform distribution between 0 and the *collision window* size. The collision window doubles after every collision, until it hits a predefined maximum, and it is reset after a successful transmission (see Figure 7.10). If at this time the station still does not hear any transmissions, it proceeds to transmit (the transmission itself may be preceded by an RTS/CTS exchange that we will explain shortly). In both cases, if the packet is a directed transmission towards one other station (rather than a broadcast), it waits to receive a positive ACK from the intended receiver. This sequence of actions helps keep the number of collisions low while allowing the station to schedule a retransmission in case a collision does occur (in which case it would not receive the positive ACK).

We now examine the carrier-sense mechanisms in more detail. Since there may in general be hidden terminals (as explained in Chapter 3), it helps to have a good carrier sense mechanism that can prevent collisions even when there are hidden terminals. Hence, 802.11 defines both physical and virtual carrier-sensing mechanisms, where the virtual carrier-sensing mechanisms make carrier sensing more robust and help in solving the hidden-terminal problem. The physical carrier–sensing mechanism is analogous to that used in Ethernet, where a station listens to find out if the medium is idle before transmitting. It is physically layer specific, but all 802.11 physical layers need to provide this service to the MAC layer. It will not detect hidden terminals, but at least can detect transmissions from other stations that an 802.11 station can directly hear. The virtual-sensing mechanisms include an RTS/CTS scheme, as well the Duration ID field in 802.11-directed frames (i.e., frames directed to a particular station's address).

The RTS/CTS mechanism works as follows: Before sending data (assuming that the physical carrier–sense mechanism does not hear any ongoing transmissions from other stations), a station A transmits an RTS frame to the intended recipient of the

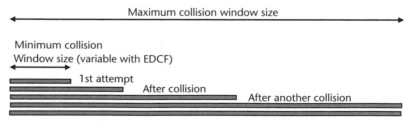

Figure 7.10 Binary exponential growth of the random back-off (drawn from uniform distributions over larger and larger intervals).

data, B. B transmits back a CTS frame. Now, all other stations within the transmission range of *both* A and B know that the transmission medium will be busy, and so they can avoid colliding with the transmission from A to B. This solves the hidden-terminal problem, as shown in Figure 7.11 (where C is no longer a hidden terminal to A, unlike in Figure 3.8). The RTS and CTS frames indicate not only that a transmission is about to occur, but the Duration ID field in these frames indicates how long the intended data transmission will take. Other stations will treat the RTS/CTS exchange as a reservation of the medium for the specified duration, and they maintain a counter, the network allocation vector (NAV), based on this information. In addition to reserving the medium for the impending transmission, the RTS/CTS mechanism also serves as a quick collision detection that checks if the transmission path is clear. Although RTS/CTS can be useful, it is clearly overhead. Hence, each station can independently decide whether or not to use it at all, or whether to use it only for data frames exceeding a certain size.

7.4.1.2   PCF

With DCF, all stations are equal, in the sense that they all contend for the use of the medium, without special privileges. While this is fair, a little unfairness is sometimes desirable, for example, to provide QoS. Thus, when there is an AP, 802.11 provides for the AP to control the use of the medium through the PCF. In other words, during the CFPs, the stations no longer contend for the medium, but transmit only as allowed by the AP. The AP can seize control of the medium because of the differential interframe-spacing scheme in 802.11. This is the major piece of CSMA/CA that we did not discuss in Section 7.4.1.1, so we turn our attention to it now.

The standard 802.11 defines four interframe spacings, namely, short interframe space (SIFS), PCF interframe space (PIFS), DCF interframe space (DIFS), and extended interframe space (EIFS), from the shortest to the longest. In different

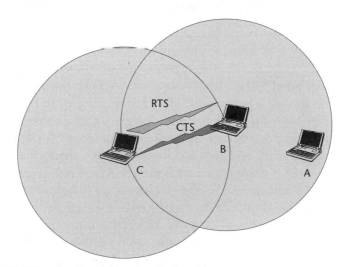

Figure 7.11   RTS/CTS solves the hidden-terminal problem.

situations, a station needs to wait for the medium to be clear for one of these inter-frame spacings, before transmitting (and in some cases, it must wait an additional time interval, based on the independent back-off mechanism). The differences in the timings of these interframe spacings allow for: (1) better collision avoidance, (2) priority for MAC control frames, and (3) control of the medium for PCF purposes. In Section 7.4.1.1, we explain how when the medium is busy when a station wishes to transmit, it needs to wait for DIFS before waiting for the random back-off to transmit. Actually, to be more precise, if it also turns out that the last transmitted frame detected was in error, the station waits for EIFS instead, improving collision avoidance. Meanwhile, SIFS, the shortest of the interframe spacings, are used to rightfully give priority to certain MAC control frames, such as ACK and CTS frames. After SIFS, PIFS are the next shortest time. Thus, an AP can, and does, wait only for PIFS to transmit a special beacon frame indicating the start of the CFP.

As soon as the other stations receive the beacon frame from the AP at the start of the CFP, they stop contending for the medium for the duration of the CFP. The duration, and other parameters, are announced by the point coordinator (the AP), in the beacon. Instead of contending for the medium, the stations are polled individually by the point coordinator. The receipt of a poll is an indication to a station that it may transmit a frame, if it has a frame to transmit. Otherwise, it responds to the poll with a null frame. It then waits to be polled again, or until the CFP ends, before it can transmit again. Thus, in controlling the sequence of polling, the point coordinator controls the use of the medium by the stations, during the CFP.

We now return to the Duration ID field in packets (including RTS/CTS frames). We have already seen that the Duration ID field is used in RTS/CTS frames to indicate the length of the intended data transmission referred to by the RTS/CTS exchange. It is also found in regular data frames, where in a contention period it indicates how long it would take for the frame to be transmitted, followed by an ACK, and one SIFS interval. This allows other stations to compute their NAV accordingly. In a CFP, on the other hand, the Duration ID field is essentially unused (it is set to a fixed constant).

7.4.1.3   802.11e

While the original 802.11 MAC incorporates a priority scheme, with the differential interframe spacings, it is restricted to giving priority to certain control frames. All data frames have the same priority. This is unacceptable for certain applications, including the important case of voice and video over WLAN.

Therefore, IEEE has been working on QoS enhancements to 802.11 that support differential treatment of data frames and facilitate voice over WLAN. The standard 802.11e introduces the new coordination functions, the enhanced distributed coordination function (EDCF) and the hybrid coordination function (HCF), to provide support for eight traffic classes. The EDCF is designed as an enhancement of the DCF built upon the DCF mechanisms, so non-QoS-enabled stations can coexist with QoS-enabled stations. It introduces traffic class-dependent interframe spacings and traffic class-dependent minimum initial collision window sizes. Instead of all traffic waiting at least DIFS after the medium is sensed free, they wait a new arbitration interframe space (AIFS), which is shorter for higher-priority traffic and longer for lower-priority traffic. As with DCF, there is a collision window. However, with EDCF, the minimum size can differ, depending on the traffic class, so even this probabilistic component can be skewed in favor of higher-priority traffic.

While the EDCF is an enhancement of the DCF, the HCF can be thought of as an enhancement of the PCF. You may wonder if the point coordinator could be a natural administrator of QoS differentiation, through careful thought to the sequencing of the polling (e.g., if some stations are known to be in voice over IP sessions, to poll these stations at regular intervals to avoid excessive delay, and so forth). Indeed, the polling sequencing is not specified in 802.11, so this sounds promising at first. However, there are several problems that are found upon closer inspection (and therefore, the HCF is needed to solve these problems). First, the point coordinator (the AP) does not know what kind of traffic each station wishes to communicate. Second, the point coordinator does not know the queue lengths in each station, and moreover there was no standardized separation of traffic into classes in each station before 802.11e came along and introduced the notion of traffic classes. Third, the CFP has to alternate with the CP, so CFP cannot be in effect continuously, which may lead to problems for streaming or conversational traffic if large delays are experienced in the CPs. Fourth, when a station is polled in a CFP, it has the right to transmit as large a packet as it wishes, up to the 802.11-specified maximum packet size of 2,304 bytes that applies for any packet passed to the 802.11 MAC, not just responses to polls. It would be useful if the poll could specify a more limiting maximum packet size allowed.

Indeed, the HCF is provided to exploit the polling capabilities of the point coordinator (called hybrid coordinator, if it uses HCF, but like the point coordinator, this would generally be in an AP). The standard 802.11e enhances the 802.11 MAC so that careful polling is a viable means of providing QoS differentiation. The problems described in the previous paragraph are addressed in 802.11e. With 802.11e, information on the traffic from each station is provided by the station to the hybrid coordinator, using the QoS control field that is introduced by 802.11e. This information is detailed enough to allow a station to specify the queue size for a specific traffic class, as well as for the hybrid coordinator to specify a limit to the size of the packet that may be transmitted in response to the poll. Another advantage of the hybrid coordinator over the point coordinator is that it can initiate HCF access even during the CP, which may be useful for streaming or conversational traffic. Thus, the hybrid coordinator has a number of advantages over the original point coordinator of 802.11, in being used as a tool for providing QoS differentiation.

7.4.2   QoS and Mobility

RSVP was designed with the implicit assumption that hosts do not move and do not change IP addresses while resources reserved using RSVP are being utilized.

However, this assumption is no longer valid in IP networks that support mobility. When Mobile IP is used, packets from an MH come with source IP addresses that change, as the MH moves from subnet to subnet and uses different COAs. When the MH changes COA, the resource reservations made with the old COA become disconnected from the MH with its new COA (the RSVP-capable routers associate the reserved resources only with the old COA). Thus, after each move to a new subnet, new resource reservations need to be made, associated with the new COA. Thus, there is a disruption of service quality with each handoff, lasting longer than the disruption in IP connectivity that happens at each handoff, because only after IP connectivity is restored with the new COA can the RSVP signaling begin.

This problem of interworking RSVP and Mobile IP can be quite serious, especially if handoffs are very frequent. A number of different solutions to this problem have therefore been proposed. Many of these solutions can be broadly divided into a few categories: (1) those solutions where RSVP is modified so that it uses a different unique identifier for a reservation (instead of using the changing IP address), where the unique identifier may be the home address of the MH, or even a completely new identifier introduced for this purpose, (2) those solutions where an existing RSVP reservation can be modified to work with the new IP address, after each handoff; for example, the resources in the intermediate routers shared by the old and new path would continue to be reserved and not need to be reserved again, and (3) those solutions where new reservations are prearranged, based on expectations of where the MH may move, to avoid the long delays in re-establishing RSVP reservations. Solutions based on these, and other approaches, have been suggested [14]. Nevertheless, there is currently no clear general convergence to a preferred solution in the industry or in the IETF, although the SEAMOBY working group in the IETF is working on a generalization of the problem.

We can generalize the problem of interworking RSVP and Mobile IP in the following way: Mobile IP only ensures that during a handoff, while IP connectivity is restored after a brief interruption. However, the network provides more than just IP connectivity. Mobile IP does not transfer context state information between the old and new points of attachment to the network. Context state information is useful for support of features like QoS, as well as authentication, authorization, and accounting (AAA) and header compression. In the case of RSVP and resource reservation, unfortunately, the resource reservation state information resides in some routers that may not be mobility-aware. The SEAMOBY working group in the IETF is involved in solving the more general problem of transferring context state information during handoff. An experimental context transfer protocol is being developed [15].

7.5 Summary

As introduced in this chapter, techniques are needed for providing differentiated QoS because limited resources in networks are shared, and because the network needs to be able to treat different classes of traffic differently based on issues including different application requirements and different subscriptions. We overview a number of different mechanisms for IP QoS, including resource reservation (using RSVP), admission control (e.g., using bandwidth brokers), packet classification and marking (e.g., using bits in the ToS field of the IP header), queuing (FIFO, priority queuing, WFQ, and CBQ), traffic shaping (leaky bucket and token bucket), policing (e.g., policing token bucket), and traffic engineering or QoS routing. We touch upon the two QoS frameworks for IP networks, the IntServ and DiffServ frameworks that provide ways to systematically apply QoS mechanisms. Then we explore QoS in wireless networks, looking at some length at QoS in 802.11, and the interactions of mobility protocols with QoS mechanisms.

References

[1] Braden, R., D. Clark, and S. Shenker, "Integrated Services in the Internet Architecture: An Overview," RFC 1633, June 1994.

[2] 3GPP TS 23.107, "Quality of Service (QoS) Concept and Architecture (Release 5)," December 2003.

[3] Braden, R., et al., "Resource ReSerVation Protocol (RSVP)—Version 1 Functional Specification," RFC 2205, September 1997.

[4] Nichols, K., V. Jacobson, and L. Zhang, "A Two-Bit Differentiated Services Architecture for the Internet," RFC 2638, July 1999; formerly draft-nichols-diff-svc-arch-00.txt.

[5] Durham, D., et al., "The COPS (Common Open Policy Service) Protocol," RFC 2748, January 2000.

[6] Ferguson, P., and G. Huston, *Quality of Service: Delivering QoS on the Internet and in Corporate Networks,* New York: John Wiley & Sons, 1998.

[7] Blake, S., et al., "An Architecture for Differentiated Services," RFC 2475, December 1998.

[8] Shenker, S., C. Partridge, and R. Guerin, "Specification of Guaranteed Quality of Service," RFC 2212, September 1997.

[9] Ghanwani, A., et al., "Traffic Engineering Standards in IP Networks Using MPLS," *IEEE Communications Magazine,* December 1999.

[10] Wroclawski, J., "Specification of the Controlled-Load Network Element Service," RFC 2211, September 1997.

[11] Heinanen, J., et al., "Assured Forwarding PHB Group," RFC 2597, June 1999.

[12] Jacobsen, V., K. Nichols, and K. Poduri, "An Expedited Forwarding PHB," RFC 2598, June 1999.

[13] Grossman, D., "New Terminology and Clarifications for Diffserv," RFC 3260, April 2002.

[14] Paskalis, S., and A. Kaloxylos, "An Efficient RSVP-Mobile IP Interworking Scheme," *Kluwer Mobile Networks and Applications,* June 2003, pp. 197—207.

[15] Loughney, J., et al., "Context Transfer Protocol," draft-ietf-seamoby-ctp-08.txt, January 2004, work in progress.

CHAPTER 8

Network Security

Security and QoS are two-thirds of a triumvirate, the third part of which is mobility. They are the major network-level areas of technical challenge that wireless IP must successfully handle to succeed. Of the triumvirate, security and QoS are more similar to each other than to mobility. Whereas mobility is for the most part an interesting problem only in the wireless case, security and QoS are among the most important areas of work and development for IP, whether wired or wireless. In both cases, there are challenging and interesting problems in the wired side, with additional challenges when wireless is considered.

One reason why security is so challenging is that it is a negative problem [1]. In many other problems in the wireless IP world, the objectives are to achieve something that can be verified with reasonable effort. For example, it is easy to verify that the objectives of a protocol like Mobile IP are met, when packets get forwarded to the correct locations of MHs. However, it is difficult to verify that a network is secure, because we would need to test it against a variety of different attacks. Even then, vulnerabilities might not be spotted, only to be revealed by yet another type of attack that the designer has not considered. Furthermore, in many other cases, a natural feedback mechanism exists when something fails—the user can be expected to complain. However, if security fails, the person or persons with knowledge of the failure are typically the attackers, who would often not have an incentive to report the security failure.

8.1 Introduction

We consider only network security here. Thus, we examine only security issues related to communicating over a network. This includes security issues related to communicating over a wired or wireless data link, but not machine security. Machine security includes proper and careful design of operating system software and machine hardware to avoid security holes that can be exploited by malicious users. It also includes physical security, such as keeping machines in secure locked areas, and proper care of laptops in places like airports where some thieves target laptops and PDAs. A practical system designed with security in mind would need to consider machine security and network security, as well as carefully designed procedures for the human operators (e.g., to avoid a malicious user stealing an authorized operator's key and accessing the system).

8.1.1 Requirements

Imagine two kings at war, allies in a fight against a common enemy. Suppose the two allied kings are in two different camps. They need to communicate with each other and with their generals, and security is important because the battle could be lost if their communications are compromised (e.g., if their enemy manages to read their messages to each other). In particular, messages sent from one king to another need to be private and confidential. Hence, the first requirement is the *privacy* or *confidentiality* requirement, that somebody who somehow intercepts the messages should not be able to understand them. However, an interceptor may alter or corrupt the message (this may happen without malicious intent if, for example, the loyal messenger memorizes the message but inadvertently forgets part of it). A second requirement, then, is assurance that the message has not been altered—this is known as the *message integrity* or *data integrity* requirement.

What else could the enemy do? In other words, what other *attacks* could the enemy launch? The enemy could send a fake message purportedly from one king to the other. Therefore a way to ensure that message is really from whom it claims to be (i.e., to authenticate the message) is needed. This is the *message authentication* requirement. Whereas message authentication ensures that a message is from whom it claims to be, *user authentication* ensures that a node (a user) is who it claims to be for network access control purposes.

Closely related to authentication is *authorization*. While authentication is about identity, authorization is about what that identity is allowed (authorized) to do. For example, only a king and not a general is allowed to command the use of a secret weapons cache. If a command to use this cache comes from a general, he is not authorized to give that command, even if the command can be authenticated as his. In fact, in the context of network access, authentication, authorization, and accounting are so often closely tied together, that the term AAA (authentication, authorization, and accounting) was coined for protocols like RADIUS that facilitate all three.

Confidentiality, data integrity, and authentication are arguably the big three issues in network security. Given the limited space in this chapter, we focus on the big three and briefly discuss other possible network security requirements. Attacks like the classic man-in-the-middle, spoofing, eavesdropping, jamming, and code-breaking should fail if the network is carefully designed to provide sufficient confidentiality, data integrity, and authentication. The man-in-the-middle scenario is where an attacker gets between two communicating entities, say A and B, and to A pretends to be B, and to B pretends to be A. Spoofing is pretending to be someone the attacker is not and sending messages as that other entity (so the man-in-the-middle scenario is a special case of spoofing). Eavesdropping is "listening in" on communications the attacker is not authorized to listen to. This includes unauthorized wiretaps and listening to wireless signals. Jamming is disrupting communications by introducing interference, such as electromagnetic interference, into the communications medium.

What are some other requirements? A network should also be required to not be vulnerable to *denial of service (DOS) attacks*. A DOS attack is one in which the attacker starves the target(s) of resources, typically by making many spurious requests and thus tying up resources to the point that others get starved. The classic example is where an attacker repeatedly opens TCP connections at a machine, leaving the connections half-open and thus tying up resources so legitimate TCP connection attempts are denied service. In the wireless case, *jamming*, the deliberate introduction of electromagnetic noise to disrupt communications, is considered a form of DOS attack. The requirement to be unsusceptible to DOS attacks can also be expressed as a requirement of *service availability*. The network protocol designer should also guard against *replay attacks*. These are when an attacker, unable to decipher an eavesdropped message, nevertheless stores the raw message and later replays it without change. We can express the corresponding requirement as a requirement on uniqueness—each message is unique and cannot be replayed. Common security requirements and typical solutions are listed in Table 8.1.

Table 8.1   Security Requirements

Requirements	Needed to Protect Against These Attacks	Solutions	
Confidentiality/ privacy	Eavesdropping, man-in-the-middle	Encryption schemes to make the data unintelligible to attacker	Message-specific requirements
Data integrity	Corruption of data	MAC, checksums to guarantee that the data has not been corrupted	
Message Authentication	Man-in-the-middle, spoofing, replay	Authentication schemes to guarantee the message is from who it claims to be from	
Non-repudiation	Denial of responsibility for contents of message	Digital signatures that only one sender could have used	
Uniqueness	Replay	Time stamping so the message can be recognized as invalid if replayed later; unique identifiers that are never reused (nonces) can also be used	
User authentication and authorization for access control	Man-in-the-middle, spoofing, replay	Challenge/response schemes	Network- and node-specific requirements
Service availability	Denial of service (DOS), jamming	Various schemes to reduce resource consumption by attackers	
Intrusion detection	Most attacks to "break in" to a system leave traces— intrusion detection is a "second-line" of defense to detect suspicious activity		

8.1.2   Solutions

In this chapter, we do not have the time to go into the details of the various algorithms and schemes (security mechanisms) that have been invented over the years to provide security services. Instead, in this section we briefly survey the range of mechanisms available for confidentiality, data integrity, and message authentication, and then move on to focus on IP security and wireless security, where we will see how some of these mechanisms are incorporated into security frameworks to meet the security needs. You may refer to Stallings for more details on security mechanisms [2].

To provide privacy and confidentiality, there are many encryption schemes from which to choose. There are two main classes of encryption schemes: secret key and public key. Secret key schemes are described as "conventional," "shared key," "shared secret," and "symmetric," whereas public key schemes are described as "public key," "public/private key," and "asymmetric." Secret key schemes were the only schemes around until the mid-1970s, when public key schemes were first published (hence, secret key schemes are "conventional" schemes). Secret key schemes use secret (private) keys that are shared by sender and receiver, that cannot be revealed without compromising message privacy. Decryption is symmetric to encryption in the sense that the reverse of the encryption algorithm is performed, using the shared key. These fundamentals of secret key encryption are shown in Figure 8.1. On the other hand, public key schemes are asymmetric in that they use one key for encryption and another key for decryption. Each node generates a public and private key pair so that the public key can be made known to anybody without placing the private key in jeopardy (even with knowledge of the encryption and decryption algorithms and samples of encrypted text). Then, any other node that wishes to send a message to, say, node A will use node A's public key to encrypt the message. Node A's private key is needed to decrypt the message. Since only node A has its private key, the message is kept private. Figure 8.2 illustrates public key encryption.

To provide confidentiality, there is a wide selection of encryption schemes to choose from. Examples of private/secret key encryption schemes include data encryption standard (DES), triple data encryption standard (3DES), international data encryption algorithm (IDEA), Blowfish, RC4, and RC5. Examples of public

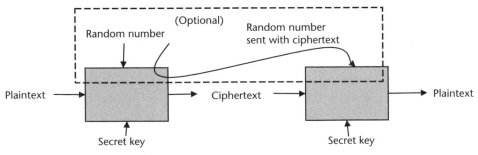

Figure 8.1   Secret cryptography.

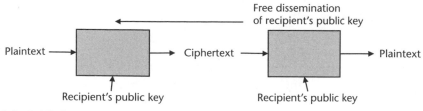

Figure 8.2   Public-key cryptography.

key encryption schemes include Rivest, Shamir, and Adleman (RSA) and elliptic curve-based cryptography. Typically, unencrypted text is called plaintext, while encrypted text is called ciphertext.

As for data integrity and message authentication, many security schemes provide both of these services together. Many schemes have been proposed, which can be classified as falling into the class of *message authentication codes* (MACs) or the class of *hash functions*. A MAC[1] is a cryptographic checksum of a message, where both the sender and receiver share a secret key. Typically shorter than the original message itself, the MAC is sent along with the original message. Like ordinary checksums, the MAC allows the receiver to detect changes in the message (if the MAC computed at the receiver does not match the MAC in the sent message). Furthermore, unlike with an ordinary checksum, an attacker cannot change the message and then alter the MAC to match, without possessing the secret key. Hence, data integrity is provided. Since only the sender and receiver have the secret key, the presence of the MAC indicates to the receiver that the message is indeed from the sender, thus providing message authentication. With hash functions, on the other hand, the sender and receiver do not need to share a secret key. Given a message, the hash function produces a hash (of the message) also known as a *message digest,* similar in some ways to a MAC in that it is shorter than the original message, but different in that the sender and receiver do not need to share a secret key. Hash functions are chosen so that it is easy to compute the hash given a message, but not the other way around (i.e., not given a hash to compute the message), and it is very hard to find two messages that result in the same hash. The second property is crucial to ensure that when a message is changed, the hash of the changed message will no longer match the transmitted hash, and so the change can be detected and data integrity can be provided. However, because there is no secret key used for computing the hash, the hash portion of the transmitted message is usually encrypted. Thus, the combination of encryption and decryption and the hash function provide data integrity and message authentication.

Examples of MACs include the data authentication algorithm based on DES and HMAC [3]. Examples of hash functions include MD4, MD5, and SHA-1.

---

1.   Unfortunately, the acronym MAC is also used in networking to refer to the medium access control layer of the protocol stack. In most cases, it should be clear which acronym expansion for MAC to use, based on the context of usage.

8.2 IP Security

The Internet provides only best-effort delivery service, just as it provides only inse-cure packet delivery, with no guarantees that packets are unread and unmodified by intermediate nodes or that packets are from whom they claim. Certain applications might be content with insecure packet delivery, such as casual Web browsing for topics of little or no consequence. However, with the growth of the Internet and especially of e-commercial and online banking, users increasingly need secure com-munications for certain kinds of applications, such as those involving financial transactions.

So there is a need for Internet users, and more generally, users of IP-based net-works, for secure communications. But first, we must ask ourselves whether it makes sense at all to put security mechanisms in the network layer, in IP in particu-lar. We compare putting security mechanisms in the network layer to putting secu-rity mechanisms in the application layer. At one level, the difference is between generality and specificity. All traffic is passed down to the network layer, so for security mechanisms at the network layer, blanket security protection is applied, whereas security mechanisms at the application layer are specific to particular applications. Specificity allows the security scheme to be tailored to the particular application, such as being much more secure for a financial transaction application than for a social chatting application and probably requiring certain services such as non-repudiation. However, one of the drawbacks of this kind of specificity is that every application has to take care to secure its own communications, and this may be a burden to some applications. Sometimes, such as in a virtual private network (VPN) scenario (to be discussed soon), *all traffic* should be encrypted, and thus it makes much sense to implement VPNs at the IP level. Thus, it is not a question of which is better, but for what applications, under what circumstances, it would make more sense using security mechanisms at the network layer or higher up.

Another consideration is that security mechanisms at the IP layer can encrypt more of the headers than can security mechanisms at higher layers of the protocol stack. This is due to how layered networking works. At the source, as a packet is passed down the protocol stack for processing, a given layer will only see the headers added by layers above it and will not see lower-layer headers. Application-layer security mechanisms, on the other hand, would therefore leave all the lower-layer headers in the clear. Having some headers exposed may make the system vulnerable to *traffic analysis*. Traffic analysis of in-the-clear headers can reveal potentially sen-sitive information to an attacker, information such as network addresses and topol-ogy and traffic loads from various sources.

8.2.1 The Need for Security

In Chapter 1, while discussing technological advances that will make the next gen-eration of wireless IP networks possible, we stress the important requirement of keeping the system at a reasonable cost. Thus, the cost of providing any given set of features must always be considered, sometimes leading to tradeoffs and compromises. A VPN is a good example of this. In many situations, organizations like to have their own private network to conduct their business. Such a network would be connected only to trusted machines within the organization. Provided that reasonable care is taken to physically secure the network equipment, including wires, the organization can be reasonably confident that outsiders cannot see what goes through the network. Hence, the network is a true private network providing privacy (in the sense of confidentiality). While this network would serve the purposes of the organization, it would be much too expensive, especially if the organization is distributed over many locations far around the world, as shown in Figure 8.3. Using a public network like the Internet, as shown in Figure 8.4, would be much more cost effective. However, it would not provide the communications privacy that the organization needs.

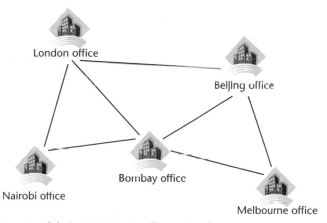

Figure 8.3   Private network between remote office networks.

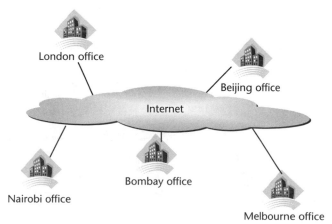

Figure 8.4   Replacing the private network with a public network.

The solution is to still use the Internet, but to construct a virtual network over it. This can be done, for example, by placing gateways at each point where the organization's network borders the Internet, and privacy can be achieved if traffic is encrypted between the gateways. Such a network then functions as a VPN, which is very popular as a cost-effective way to achieve privacy without actually needing to build a private network. It is shown in Figure 8.5, where the thick curvy lines represent secure *tunnels* (such as can be provided by IPsec, as we will see soon) between the various locations.

We have coined the term wireless link band-aid to refer to the case where a wireless link is insecure, or provides very weak security, but the wired portion of the communications path may be secure enough. For example, the original 802.11, whose wired equivalent privacy (WEP; see Section 8.3.1) is notoriously breakable. In such a case, the best long-term solution may be to change the wireless link itself, as is being done with 802.11i, and in proprietary solutions that vendors have been deploying in their products in the interim. However, an intermediate solution is to apply higher-layer security mechanisms to bear upon the problem before 802.11i is stable and widely deployed. It makes sense for this to be done at the IP layer because it is closest to the effect of a true link-layer solution without actually being one. All traffic is protected without the need to modify all the applications, and most forms of traffic analysis are protected against, since IP layer and higher-layer headers can be encrypted, unlike the provisions of an application-layer solution. Like a band-aid, it provides a temporary solution for protection while the long-term solution is developing (if we think of 802.11i development and deployment as analogous to skin regrowth). The solution is close to the skin, in the sense of the IP layer being the layer just above the link layer, and providing much of the protection that a link-layer scheme could provide. We could extend IP-VPNs to include wireless links, as shown in Figure 8.6. Table 8.2 summarizes the example applications for IP security discussed in this section.

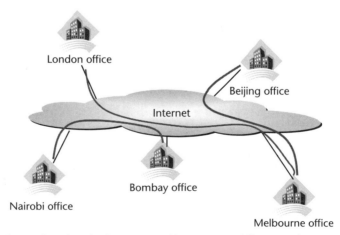

Figure 8.5   Implementing virtual private networking over a public network.

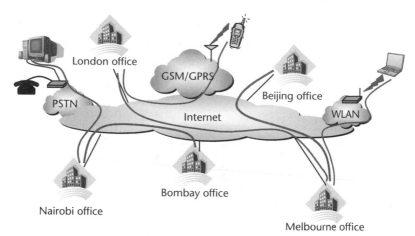

Figure 8.6   Adding support for wireless and dial-up access to the VPN.

Table 8.2   Example Applications and Needs

Applications	Needs
VPN	Create a virtual network over a shared IP network like the Internet, where all traffic must be confidential
Wireless link band-aid	Add an "artificial skin" layer of protection for all IP packets going over a vulnerable wireless link

8.2.2   IPsec

IPsec is a security framework for IP-layer security services in IP networks. IPsec provides two modes of usage, namely transport mode and tunnel mode. Furthermore, each of the modes has three submodes, namely authentication header (AH) only, encrypted security payload (ESP) only, and both AH and ESP. Using ESP provides confidentiality, where a variety of secret key encryption schemes may be utilized. Using AH provides message authentication and data integrity.

The difference between transport mode and tunnel mode is that in transport mode, the encryption and authentication operations are over only the payload, and not the IP header, whereas in tunnel mode, the entire IP packet *is* included in the encryption and authentication operations. In tunnel mode, since the IP header may be encrypted, and thus no longer usable for routing, a new, unencrypted IP header is attached. The difference between transport and tunnel modes is shown in Figure 8.7, where the shading represents the parts of the packet that are encrypted. Clearly, tunnel mode adds a little more overhead than transport mode does, because of the use of a new IP header. In the case of usage of IPsec between two end hosts, it therefore makes sense to use transport mode. Even with tunnel mode, the source and destination IP addresses of the end hosts are revealed anyway, in this case. In the case of a VPN scenario, however, if IPsec is used only between two gateways, and traffic between two end hosts goes through the two gateways along the path

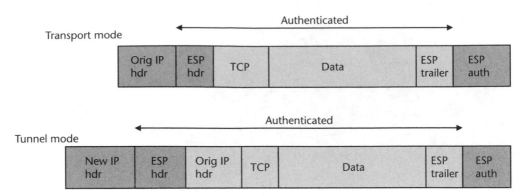

Figure 8.7 IPsec transport and tunnel modes.

between the two end hosts, then tunnel mode is more appealing. This is because with tunnel mode in this case, the original IP header is hidden, and only the IP addresses of the gateways are revealed.

8.3   Security in Wireless Networks

Security does not come for free. It must be provided for, at the cost of such things as bandwidth, power consumption, and delays. There are trade-offs involved, depending on the levels of security required for different situations. These trade-offs are better understood for wireline networks than wireless networks, given that wireline networks have been around for longer. When wireless links are involved, constraints like limited bandwidth, power, and processing capability may mean that the optimal points in the tradeoff curves change.

In some cases, wireless-specific algorithms and protocols may be beneficial. For example, wireless transport layer security (WTLS) is a version of transport layer security (TLS) optimized for the wireless environment, allowing less computationally intensive algorithms to be used at the wireless terminal and lower bandwidth overhead and faster handshakes, at the cost of a lower level of security [4]. The introduction of specialized protocols like WTLS for wireless access, however, introduces another challenge. It places a burden on servers unless there is a translation gateway that shields the servers from having to understand all the specialized protocols. The WAP gateway (see Chapter 10) is an example of one such gateway that has WTLS/TLS translation among its features [5].

One of the most important features that wireless access enables is mobility. To support mobility everywhere, the network must augment its routing, QoS, security, and other features. First, user authentication for network access when a subscriber accesses the network over a wireless link (even in a home network) is important, since the terminals are not permanently attached by wire to any given location. Second, when the subscriber is roaming, a number of challenges arise: (1) how to work with mobility protocols to support seamless provisioning of security services while

roaming and for interadministrative domain mobility, and (2) b
between the visited and home network in AAA, as well as provid
mechanisms.

8.3.1   WLAN Security

When 802.11 WLANs were first designed, it was noted that they should have the same level of security as typical LANs like Ethernet LANs. Ethernet does not encrypt traffic, but because Ethernet machines are connected to one another by cables, it is harder to tap into the LAN than for a WLAN (where an attacker need not physically connect any hardware to the LAN to eavesdrop). Therefore, the wired equivalent privacy (WEP) scheme was proposed. WEP is a weak, 40-bit secret key (symmetric) encryption scheme that was not designed to be super-strong, but to raise the difficulty of eavesdropping WLAN traffic to an equivalent level as with a wired LAN. Additionally, an integrity check value (ICV) is computed on the plaintext and appended to the plaintext before WEP encryption. Figure 8.8 shows WEP encryption; since it is a symmetric scheme, decryption is straightforward, given the secret key.

For WEP encryption, the plaintext is bitwise exclusive-ORed (in exclusive-OR is a binary operation often abbreviated as XOR, where the output is 0 if the two input bits are the same and 1 otherwise) with the output of the WEP pseudorandom number generator to produce the ciphertext. Thus, the ciphertext can be decrypted only if the output of the WEP pseudorandom number generator is known. The idea is that only the intended receiver would know the two inputs to the WEP pseudorandom number generator that the transmitter uses. These two inputs are a secret key and initialization vector (IV). How does the receiver know these values? The 802.11 standard assumes, but does not specify, the existence of an external key distribution mechanism that distributes the secret key to a set of authorized mobile stations. The IV, on the other hand, is not distributed beforehand. Furthermore, it may be changed by the transmitter as often as with the transmission of every 802.11 packet. However, the current IV value is always appended *in plaintext* to every 802.11 packet, so the intended receiver only needs to know the secret key beforehand. Note that this also means that the other stations to which the same secret key has been distributed can also decrypt the packets for the receiving station. This is

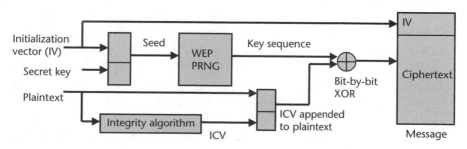

Figure 8.8   WEP encryption.

not necessarily cause for alarm, since WEP is only meant to provide privacy equivalent to that of a WLAN. Furthermore, the set of authorized mobile stations with the same secret key can be thought of as analogous to stations on an Ethernet LAN, which can hear all the traffic on the same LAN.

Nevertheless, the various weaknesses of WEP have been highly publicized [6]. As a result, an enhanced version of WEP, sometimes known as WEP2, has been developed, that uses 104-bit keys instead of 40-bit keys. Meanwhile, the IEEE came up with 802.11i, for enhancing WLAN security. 802.11i introduces a new security framework for the 802.11 family that incorporates 802.1x (for integration into the AAA infrastructure and thus providing network access control for roaming mobile stations), as well as stronger encryption algorithms. These new encryption algorithms include TKIP and advanced encryption standard (AES). Use of TKIP is meant to be an interim measure, whereas AES is a longer-term solution. The 802.11i standard closes many of the holes found with WEP.

Meanwhile, some organizations that are more security-conscious have opted for network-layer solutions to complement the security mechanisms provided at the link layer by 802.11. These solutions include the use of IPsec. In a wireless context, though, there are issues that arise from the use of both IPsec and Mobile IP together, and these will be discussed in Section 8.3.3.2.

8.3.2 GSM Security

In GSM, as with any other wireless system, user authentication is very important because users are mobile and often change their points of attachment to the network. The authentication forms the basis for ensuring that only authorized users obtain service from the network, and only for the services for which they are authorized. The other major security service that GSM provides is confidentiality, to prevent attackers from listening to mobile phone conversations. Confidentiality is mostly provided by encryption, with one significant exception that we will explain. The authentication and encryption mechanisms used in GSM are not independent. Instead, the key used for the ciphering (encryption) is computed as part of the authentication process. So we examine authentication first.

There are two kinds of authentication that are needed. First, because the phones are small, light, and portable, they can be easily misplaced and fall into the hands of unauthorized users. Thus, the human user of the phone must be authenticated with the phone through a password that is typically arranged in conjunction with the service provider upon signing up for service. More precisely, the authentication is with the SIM card—if you replace your SIM card in your phone with another, you need to know the password for the other SIM card. The human user only has access to all the features of the phone (including making and receiving calls) after successful authentication with the SIM card. Note that this authentication is only local on the phone—nothing goes over the air in this process. Thus, a second kind of authentication is needed, in which the GSM network authenticates the subscriber for service. More precisely, the GSM network actually authenticates the SIM card in a process that does not involve the human user, and that happens multiple times, regularly, when a mobile phone is on. To those who might wonder why the human user is not directly authenticated with the network, but indirectly through the SIM, we point out the analogy of the ubiquitous lock-and-key system used to safeguard most people's houses. The key is analogous to the SIM card, and the lock to the authentication procedure with the GSM network. The lock authenticates the key, not the human! Anybody with the right key can open the lock. What is to prevent somebody from stealing the keys and using them? The person has to know the right house in which to use the keys. Knowing the right house is like knowing the password in the user authentication to the SIM card.

Authentication of the SIM by the network is done through a challenge/response mechanism. The SIM and the authentication center (AuC) in the MS's home network share a key, K_i, which must never be revealed to third parties. K_i and a random number are used by the AuC as input to an algorithm, A3, to generate a value, signed response (SRES). The network sends the SIM a challenge, namely the random number. It expects that only the SIM with K_i can use A3 to generate the correct SRES for the input pair of K_i and the random number. Thus, the response sent back to the network is the SRES the SIM computes. The network compares the two values, and if they match, the SIM is authenticated. GSM authentication is illustrated in Figure 8.9 (note that this is a conceptual view of GSM authentication, because we have been referring to "the network" as a single entity, whereas actually two network elements are involved; a more complete picture of GSM authentication is given in Figure 8.10).

Most of the time, user traffic is encrypted for privacy. However, the ciphering (encryption) can be turned on only after authentication completes. This is because the key used by GSM encryption, K_c, is computed during authentication. Therefore, GSM cannot provide privacy for the control signaling that initiates

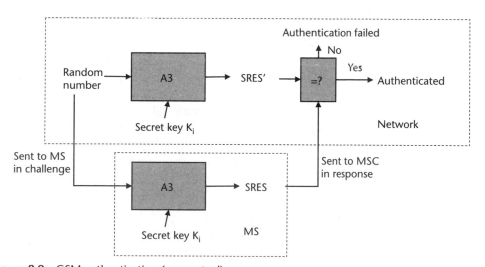

Figure 8.9   GSM authentication (conceptual).

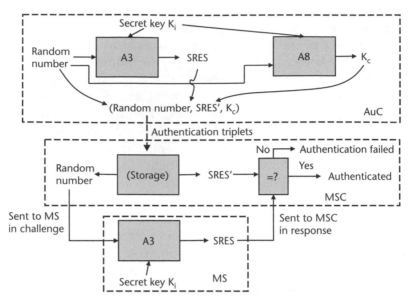

Figure 8.10 GSM authentication.

authentication. The main concern is that the MH needs to identify itself at this time. In particular, the network needs to obtain the MH's IMSI from the MH. The problem is that sending the IMSI regularly over the air unencrypted opens the door to theft of the IMSI and other privacy issues. GSM uses an ingenious workaround; instead of sending the IMSI over the air each time the MH needs to identify itself to the network, it usually sends what is known as a temporary mobile station identifier (TMSI). The link between TMSI and IMSI is known by the serving MSC, and GSM provides for a way to pass this knowledge to the next serving MSC, upon handoff to a new MSC. Thus, the sending of the IMSI itself over the air in plaintext is minimized.

What do we mean in the last two paragraphs by "the network"? It could be the AuC doing all this, but typically it is the serving MSC, so the signaling does not have to take a long time going between the AuC in the home network to the MS in a foreign network, when it is roaming. But the AuC cannot reveal the K_i to any MSC. Thus, the solution is for the AuC to precompute *authentication triplets* comprising of (random number, associated SRES computed with that random number and K_i, resulting K_c), and to send these triplets to the MSC in batches. Thus the MSC only once in a while would contact the AuC to refresh the triplets. Figure 8.10 shows the more complete picture of GSM authentication that includes the authentication triplets.

Finally, we turn to confidentiality in GSM. As we mentioned, K_c is derived during authentication. The same two inputs, K_i and the random number, are sent to a different algorithm (A8, instead of A3) to generate K_c. The encryption scheme used is a secret key scheme. Two new keys, S1 and S2, are generated for each frame, using the A5 algorithm. The inputs to A5 are the frame number and K_c, so S1 and S2 will change from frame to frame. S1 is used to XOR the traffic from the network to the MS, while S2 is used to XOR the traffic from the MS to the network.

8.3.3   Security and Mobile IP

Mobile IP provides basic support for authentication, but not confidentiality or data integrity. This is because authentication is more of a challenge where there is mobility than in networks without mobility. How does a HA know if the registration message it is receiving from a foreign network is really from one of the MHs it serves? Meanwhile, the need for confidentiality and data integrity are not necessarily increased in mobility situations. Therefore, confidentiality and data integrity are presumed to be handled by other IP protocols (as mentioned in earlier chapters, the IP protocols are modular and designed to be used in concert with other IP protocols). Since Mobile IP does not make assumptions about the security of the wireless link, it has to assume that transmissions can be heard by attackers (eavesdropping). Therefore, replay attacks may be possible, and so the authentication schemes must be protected against replay attacks.

Which entities need to have security associations with each other? At a minimum, the MH and its HA must be able to authenticate each other's messages. Otherwise, if the HA does not authenticate the registration message from the MH, any node could claim to be the MH and maliciously register an arbitrary IP address as the MH's COA. Similarly, if the registration response from the HA is not authenticated, an attacker could intercept and destroy the MH's registration message, and state falsely that the HA has updated its binding for the MH to the latest COA, by sending an unauthenticated registration reply to the MH. Therefore, Mobile IP makes it mandatory for both registration messages and registration replies to be authenticated. This is done by including the *mobile-home authentication extension* in the messages. (Mobile IP uses a general method of allowing miscellaneous, optional information to be attached to Mobile IP messages, while extensions are self-contained sequences of information, including an extension type code and length.) The mobile-home authentication extension contains a 4-byte SPI, that, together with the home IP address of the MH, uniquely identifies an MH-HA security association. The default authentication algorithm is HMAC-MD5 [3].

Additionally, there may also be security associations between the MH and FA, and between the FA and HA. Two optional authentication extensions, the mobile-foreign authentication extension and the foreign-home authentication extension, are optionally attached to Mobile IP registration messages and replies. These would only make sense when an FA is used—thus, they are not used when a colocated COA is being used. Although these two are optional, they must be used when the MH and FA, or FA and HA, respectively, share a security association. Note that of the three authentication extensions, the Mobile-Home and Mobile-Foreign Authentication Extensions (if used) are added by the MH in registration requests, whereas the Foreign-Home Authentication Extension (if used) is added by the FA in the registration request, when it forwards the registration message from the MH.

8.3.3.1 Mobile IP and AAA

We note that the original Mobile IP security model, as just described, assumes that the MH is able to actually send Mobile IP registration messages, and respond to them, soon after entering the foreign network. This implies that the foreign network has given the MH sufficient access to its network to be able to do the registration. More specifically, if colocated COAs are used, this means the MH is able to obtain an IP address in the foreign network, and if FAs are used, this means that the FA is receiving and processing Mobile IP registration messages from the MH.

How valid is this assumption? In many cases, for mobility *within an organization* (more specifically, where the points of attachments used are all part of the same network administrative domain), it is valid. For example, if the points of attachment are WLAN APs, all APs and MHs belonging to that organization may use the same WEP key. However, it turns out that the assumption is not valid in many practical situations. In many practical situations, an MH may move into areas that are part of a different network administrative domain than its home network administrative domain. This may be the case if the MH is part of an organization that is so large that it wants to have multiple administrative domains (e.g., one per office location or per business unit), or if the MH simply moves into a network belonging to a different organization. In these cases, the MH may have trouble obtaining connectivity to perform Mobile IP registration.

A good analogy offered by Perkins is that the designers of Mobile IP assumed connectivity would be provided as a courtesy service to visitors, in the same way that free electricity is provided to visitors to charge their laptops [7]. However, this turns out not to be the case. I think this connectivity is more like library borrowing privileges—not casually given to visitors! And there are good reasons for treating connectivity like library borrowing privileges rather than for treating it like electricity. Network resources are a valuable commodity, like library books—use of network resources may not infrequently affect network performance for other, more legitimate users. This is rarely the case for electricity.

Often, in these cases of movement between administrative domains, the network prevents foreign or unknown terminals from having access. Two main classes of access control are (1) access control schemes that work at the link layer and (2) access control schemes that work at the network layer. An example of a link-layer access control scheme is the use of WEP, where the user has to know the WEP key to be able to establish a wireless link. Another example is a WLAN authorization scheme that only allows establishment of link-layer connectivity with an AP if the MH has a MAC address that is in a database. With these schemes, link-layer access is denied to unauthorized network outsiders, so an MH would be unable to send *any* IP packets, not to mention Mobile IP registration messages. An example of a network-layer access control scheme, on the other hand, might be to allow wireless links to be established, but to place an access router behind the wireless link, on the network side, that controls further access to network resources. The access router may allow limited access only to an AAA server and not to other network services.

After the exchanges with the AAA server, the access router then opens up access to the relevant set of network services as appropriate.

Whether access control is of the link-layer or network-layer variety, the obvious solution is for AAA servers in the foreign network to communicate with AAA servers in the home network. This is because the foreign network AAA server probably does not have information about the MH, while the home network server should have such information. A typical arrangement is that the operators of the two networks have agreed to allow subscribers from each others' networks to access an agreed-upon set of services. This set of services may be small, perhaps just basic IP connectivity with best effort service, or it may also include other services like preferential queuing services in routers. In any case, the AAA server in the foreign network communicates with the AAA server in the home network so the AAA server in the home network can *authenticate* that the MH is indeed one of the home network's subscribers. Furthermore, since subscribers may not all have the same authorization for services, the server can *authorize* the MH for the appropriate set of services. Lastly, *accounting* can be performed so that network usage can be monitored and the subscriber billed appropriately.

A common and popular protocol for AAA is remote authentication dial-in user service (RADIUS) [8]. A more modern protocol for AAA is DIAMETER [9]. Whatever the AAA protocol used, the next question is how to get it to work with Mobile IP. The first option that might come to mind is to use AAA first, thus providing connectivity, and then to do Mobile IP registration, thus providing proper routing for the home IP address. However, this slows down the handoff process. Already, with Mobile IP alone, there can be serious handoff latency problems, and adding the latency from the AAA communications will cause more user unhappiness. Therefore, the currently in-favor model is to combine the AAA and Mobile IP signaling by piggybacking Mobile IP registration messages on AAA messages, as shown in Figure 8.11. This helps with latency, because only one round-trip is made between the two networks, rather than two, in the case that AAA is first completed before Mobile IP begins. This comparison is illustrated in Figures 8.12 and 8.13.

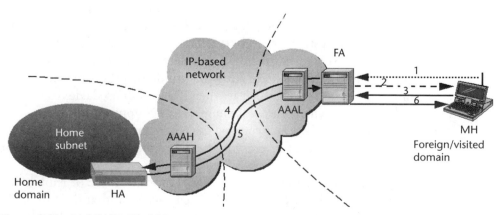

Figure 8.11   Mobile IP with AAA.

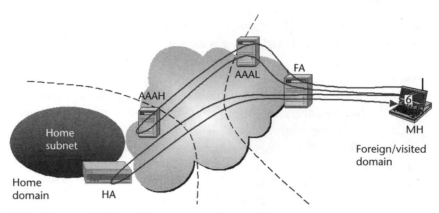

Figure 8.12 Two round-trips add latency to registration process.

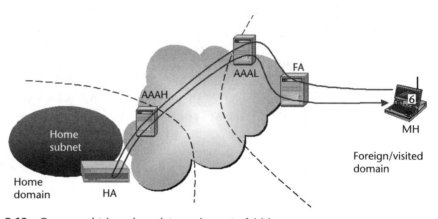

Figure 8.13 One round-trip reduces latency impact of AAA.

8.3.3.2 Mobile IP and IPsec

We have seen how the use of Mobile IP may result in problems with the use of RSVP for QoS purposes, because RSVP was designed with the implicit assumption that the IP addresses of the end hosts do not change during the time that RSVP is used. Similarly, because an IP destination address is one of the three parameters that identifies an IPsec security association, we might reasonably expect that the simultaneous use of Mobile IP and IPsec results in the same kinds of problems. Actually, in many cases, this may not be an issue, such as when IPsec is not used end to end but between two fixed gateways, which is the case in many VPN applications. However, in some applications it may become an issue, such as when one end of the IPsec tunnel is at an MH, as would be the case in the wireless band-aid application we introduced in Table 8.2.

Some proposed solutions suggest modifications of IPsec or Mobile IP to make them work together. For example, it has been suggested that the Mobile IP tunnels could be modified to be IPsec tunnels bearing either an ESP or AH (or both), instead of being IP-in-IP tunnels [10]. Another approach introduces the concept of a wireless security gateway that intercepts all packets on the wired side headed for the HA [11]. Thus, regular IPsec packets are inserted into, and removed from, the Mobile IP tunnel. Thus, there is dual encapsulation between the HA and FA. The IP addresses on both sides remain unchanged whether the MH is at home or roaming; in particular, it is the IP address of the security gateway interface that faces the HA on one side, and the home IP address of the MH on the other side (since the IPsec tunnel is outside the Mobile IP tunnel).

8.4   Summary

We look at security in this chapter. After introducing the kinds of security services typically required for network security (confidentiality, data integrity, message authentication, nonrepudiation, uniqueness, user authentication, user authorization, service availability, and intrusion detection), we briefly introduce the range of cryptographic algorithms, such as DES, at our disposal. We then consider IP network security in particular, touching upon requirements and briefly introducing IPsec. More time is spent on security in wireless networks, including WLAN security and GSM security. We explore the interactions of Mobile IP with security protocols like IPsec and AAA protocols.

References

[1]   Saltzer, J. H., and M. F. Kaashoek, "Topics in the Engineering of Computer Systems," class notes for Computer Systems Engineering course at MIT, 2003.

[2]   Stallings, W., *Cryptography and Network Security*, 2nd ed., Upper Saddle River, NJ: Prentice Hall, 1999.

[3]   Krawczyk, K., M. Bellare, and R. Canetti, "HMAC: Keyed-Hashing for Message Authentication," RFC 2104, February 1997.

[4]   Dierks, T., and C. Allen, "The TLS Protocol Version 1.0," RFC 2246, January 1999.

[5]   http://www.openmobilealliance.org/; formerly http://www.wapforum.org.

[6]   Cam-Winget, N., et al., "Security Flaws in 802.11 Data Link Protocols," *Communications of the ACM, Special Issue on Wireless Networking Security*, Vol. 46, Issue 5, May 2003, pp. 35–39.

[7]   Perkins, C., "Mobile IP Joins Forces with AAA," *IEEE Personal Communications Magazine*, August 2000, pp. 59–61.

[8]   Rigney, C., et al., "Remote Authentication Dial In User Service (RADIUS)," RFC 2865, June 2000.

[9]   Calhoun, P., et al., "Diameter Base Protocol," RFC 3588, September 2003.

[10] Binkley, J., and J. McHugh, *Secure Mobile Networking Final Report*, June 1999; http://www.cs.pdx.edu/research/SMN/final.ps.

[11] Barton, M., et al., "Integration of IP Mobility and Security for Secure Wireless Communications," *ICC 2002*, New York, June 2002.

CHAPTER 9

IPv6

The version of the Internet Protocol upon which today's Internet is based is version 4, IPv4. The next version of IP is IPv6 [1]. At a high level, the reasons for IPv6 are the changing requirements for the Internet and shortcomings in IPv4. These reasons are related in some ways; some IPv4 design choices that were appropriate at the time they were made are no longer the best choices for today and tomorrow's Internet. For example, the IPv4 header checksum is recomputed at each hop, because of the unreliability of some links in the past. This is a shortcoming for modern networks with more reliable links, where it is harder to justify the overhead involved in doing this error check at every hop. We will explore the reasons and motivations for IPv6 in more detail when we discuss the IPv6 design considerations in Section 9.1.

The basic protocols for IPv6 have been specified from as early as 1998. So one of the big questions in telecommunications is, when will IPv6 take off? Despite all of the changing requirements listed above, it is difficult to switch from a working system without a simple and compelling reason as a catalyst for change. In the case of IPv6, this reason is that the IPv4 address space is being depleted. Due to address depletion, IPv4 will no longer be viable in the near future. Estimates place the end of IPv4 somewhere in the range of 2008 to 2015 [2]. However, the uptake of IPv6 has been slow so far.

You may wonder what happened to IPv5. Did somebody miscount? Actually, there was some work that went into the design of what would have been IPv5. However, it was an experimental protocol, and it was scrapped. Subsequent work went into the next version of IP, namely IPv6, and thus IP is going straight from IPv4 to IPv6.

9.1 IPv6 Design Considerations

In Chapter 2 we discuss that as the Internet has evolved, so has the way it is used. Certain requirements have emerged or grown in significance (or can be foreseen to emerge and grow in significance). For example, with the rise of streaming multimedia and other traffic with different QoS requirements, QoS support is becoming more important. With more and more commercial usage of the Internet and the growth of wireless IP, good security is becoming increasingly necessary. The growth of wireless IP also makes IP mobility support more necessary. The need for good multicast capabilities is growing with the increase of applications, like multiparty conferencing, that can make use of it. In the early days, people were happy to get a network up and working, but today, plug-and-play network configuration is becoming increasingly appreciated. Meanwhile, in today's Internet, the benefits of hierarchical routing (e.g., aggregation, smaller routing tables) are not completely realized, for example, because of undisciplined allocations of IP addresses in the earlier days. Better hierarchy would be useful.

9.1.1 Shortcomings of IPv4

Perhaps the most serious practical problem with IPv4 is the limitations of its 32-bit address space. Not all the 2^{32} addresses are available for usage as ordinary addresses for unicast, since some are special broadcast or multicast addresses. Also, as we see in Chapter 2, the original way of allocating addresses in chunks of class A, class B, and class C addresses, meant that a lot of valid addresses were unused and wasted. Measures like the introduction of CIDR and the increasing popularity of NATs have helped with address management. However, these temporary solutions alleviate the symptoms, but do not solve the root problem, as seen in Figure 9.1.

Typically, networks based on IPv4 are manually configured by human system administrators. Protocols like boot protocol (BOOTP) and DHCP, derived from BOOTP, help to automate some of the address configuration tasks. However, DHCP requires a DHCP server, and is not available everywhere. Moreover, in certain situations, the services of a DHCP server are not available. For example, such a situation arises when a group of colleagues wishes to set up a temporary ad hoc LAN (e.g., by Ethernet or WLAN) for communications with one another, perhaps at a customer site or at a convention. IPv4 does not provide an easy way for them to automatically configure their laptops for this type of communications.

There are also some shortcomings in the way IPv4 headers are processed. These are:

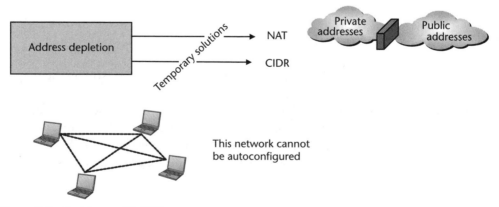

Figure 9.1 Problems with IPV4.

- Fragmentation and reassembly processing may take place in every hop between an IPv4 source and destination; this contributes to the processing effort at IP routers.
- Header checksum is checked at every hop; this contributes to the processing effort at IP routers.

We will discuss these problems, and the corresponding IPv6 improvements, when we discuss the IPv6 header.

9.1.2   Other Desirable Features

In the case of IPv6, with the much larger address space and the design to support many more nodes than with IPv4, scalability becomes especially critical. From more practical considerations, two other goals in the IPv6 design emerge. First, given the lessons learned from experiences in IPv4 network administration, and seeing that the introduction of DHCP in IPv4 is a step in the right direction, enhanced autoconfiguration features are clearly desirable. This reduces the burden on system administrators and can also be viewed as a scalability feature since host configuration becomes more scalable—system administrators can concentrate on the overall network architecture and router configurations. Second, there will not be a simple switch from IPv4 to IPv6. IPv4 and IPv6 networks will coexist for a long time and will have to interwork. This principle of the need to respect the incumbent system in the technology transition problem (as introduced in Chapter 1) is analogous to how VoIP protocols need to interwork with the PSTN because the PSTN will still be around for a long time.

9.2   IPv6 Feature Overview

The IPv6 design takes into account all the design considerations mentioned above. Highlights of IPv6 include:

- IPv6 uses 128-bit addresses, instead of the 32-bit addresses that IPv4 uses [3]; implications of the larger address space are discussed in Section 9.2.1.
- IPv6 features more powerful and flexible autoconfiguration capabilities than available with IPv4; some of these features (e.g., autoconfiguration of a link-local address) apply to both hosts and routers, while others apply just to hosts.
- IPv6 provides better mobility support than IPv4, with supporting features like autoconfiguration.
- IPv6 provides better QoS support than IPv4, with the addition of a Flow Label field in the IPv6 header.
- While not an IPv6-only feature, IPsec is expected to be more widely used in IPv6 networks than IPv4 networks.

- A revised IP header reflects the design choices made for IPv6; certain fields from the IPv4 header are no longer necessary, whereas other fields are added (e.g., Flow Label for QoS support), and others modified.
- Provisions have been made for interworking between IPv4 and IPv6, to smoothen the technology transition.

9.2.1 Implications of Larger Address Space

Clearly, IPv6 needs a larger address space than IPv4. However, having so many more addresses means much more than just being able to have enough addresses for everyone. From a system design perspective, the change from 32-bit addresses to 128-bit addresses has profound and far-reaching implications because it allows many more nodes to be on the Internet than with IPv4. Four broad classes of implications are:

- The header must be streamlined, containing as little extraneous or less-useful information as possible, because the source and destination addresses are already going to be 128 bits each.
- It makes it even more important for the design to be scalable, since there could be many more nodes than with IPv4.
- It allows IPv6 to support useful new features such as the autoconfiguration and hierarchical routing features, by subdividing the huge address space in simple ways. Some of these features are partially support in IPv4, but the huge address space allows "wastage" of IP addresses to better support these features, in a way that could not be done with IPv4 given its address space constraints. For example, a huge chunk of IPv6 addresses is used only for link local addresses (we will explain this concept shortly). Also, IPv6 supports a richer addressing scheme (e.g., for multicasting), so broadcast addresses are not needed.
- It allows every node to have at least one unique global address, so private addresses and NATs can be done away with; this alleviates all the problems that we have faced with NATs in IPv4, such as NATs breaking certain protocols.

9.2.1.1 The Streamlined IPv6 Header

The IPv6 header is shown alongside the IPv4 header in Figure 9.2. It retains the protocol version field, where IPv6 headers contain a value of 6 instead of 4. The header length field is no longer necessary, because the IPv6 header is fixed in length. The ToS field is replaced by the Traffic Class field, which has a similar purpose. Meanwhile, the identification, flags, and fragment offset fields in IPv4 are no longer needed in IPv6 because fragmentation is performed at the source (more details in Section 9.3.1). The Flow Label is new and meant for QoS support. It was added despite the need to minimize the size of the IPv6 header, because of the increasing

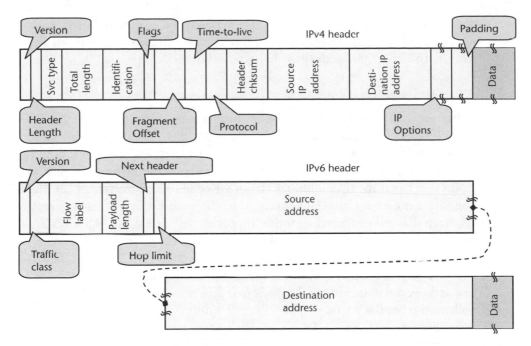

Figure 9.2   IPv4 and IPv6 headers compared.

importance of QoS support. The Payload Length in IPv6 corresponds to the Total Length in IPv4, and the Hop Limit in IPv6 corresponds to the TTL in IPv4. The Header Checksum field from IPv4 has been removed, because links are more reliable these days so the packet does not have to be checked for errors (by computing the checksum) at every hop. The protocol field is used in IPv4 to indicate the protocol header that will follow, such as a TCP header. This has been generalized and replaced by the Next Header field in IPv6, where the next header *may* be a transport layer header like TCP, or one of the IPv6 extension headers (see Section 9.3.1 for more details).

Even with the streamlining of the IPv6 header, it is still longer than the IPv4 header, mostly because of the size of the source and destination IP addresses. The IPv6 header is 40 bytes long, whereas the IPv4 header is 20 bytes long plus the length of the options.

9.2.1.2   Scalability

Several facets of scalability are incorporated in the IPv6 design. First, while we argue that the huge address space increases the need for scalability, the availability of all these addresses itself is a scalability feature, conversely, because it allows the network to scale to so many more nodes than an IPv4 network. Second, the more powerful autoconfiguration features (see Section 9.3.2) contribute to scalability by reducing administrator effort per IPv6 host. Third, the reductions in header-processing effort (see Section 9.3.1) in IPv6 help scalability. Fourth, as we will discuss next, more efficient hierarchical routing architecture in the Internet core with IPv6 (than with IPv4) helps scalability.

9.2.1.3 Hierarchical Routing

A hierarchical routing architecture has been a source of scalability since IPv4. However, blocks of IP addresses were not efficiently distributed in the past, and classless addressing arrived when there was already a shortage of addresses. Good practices of assigning large contiguous blocks to ISPs and requiring sites that wanted IP addresses to get them as subblocks from ISPs can reduce the size of routing tables in the Internet core and make hierarchical routing more efficient. However, these habits started to be practiced in today's Internet relatively late. As a result, the IPv4 Internet is somewhat fragmented, and ISPs typically have to advertise routes to many relatively smaller chunks of IP addresses.

In IPv6, hierarchical routing is made more efficient, because blocks of addresses are aggregated more efficiently, and good practices of address assignment should be followed from the beginning. Blocks of IP addresses are allocated to ISPs in the following way: the largest ISPs, analogous to tier 1 ISPs in the IPv4 world, are designated as top level aggregators (TLAs) and each assigned a large block of addresses. TLAs are at the top level of the hierarchy of address assignments. All of a TLA's customers (typically, smaller ISPs) obtain chunks of addresses from within the allocation of that TLA. The second-level ISPs are designated as next level aggregators (NLAs). Thus, a TLA should be able to aggregate routes to its NLA customers into a single address range, so fewer routes need to be passed between TLAs. Similarly, each NLA divides its block of addresses into smaller blocks for assignment to its customers, either very large organizations or smaller ISPs, known as site level aggregators (SLAs).

9.2.2 Addressing

In order to support some of its new features, IPv6 introduces an addressing concept that is richer than that of IPv4. Thus, IPv6 introduces the following ideas:

- Scope: This has to do with how widely an address is applicable, as we will see below (refer also to Figure 9.3).
- Anycast addresses: We will discuss this shortly.

For unicast addresses, there are three address scopes: link-local, site-local, and global. Link-local addresses are meant for communications on a link only, such as on a LAN. The same link-local address can be used in multiple LANs simultaneously without conflict. Routers will not forward packets with a link-local address as either the source or destination address, or advertise such routes on other interfaces, thus enforcing proper usage of link-local addresses. Site-local addresses, meanwhile, are local to sites that could include multiple LANs, such as university or corporate

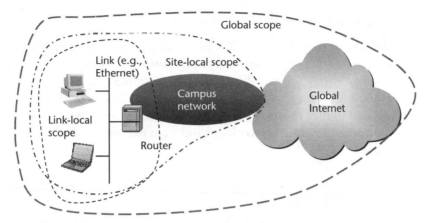

Figure 9.3   Address scopes in IPv6.

networks. Global addresses are like the normal IPv4 addresses, having global scope. The use of link-local and site-local addresses in IPv6 is analogous to the use of private addresses in IPv4.

In some cases, IP packets are not necessarily sent to one specific destination, but to any node that can provide a particular service. For example, a DNS query or a SIP server query only requires that the query reaches a node that can provide a correct response. In such cases, multiple servers may be provided, for load-balancing purposes, and they can share an anycast address. Unlike with multicasting, only one node from the group receives a packet destined to an anycast address. The node that receives it should be the one closest to the source.

Link-local, site-local, and global unicast addresses, and multicast addresses, are distinguished by different address prefixes, namely, 1111111010, 1111111011, 001, and 11111111, respectively. Many other prefixes are currently unassigned or reserved. Interestingly, the global unicast address space (since the prefix is 3 bits long) is only one-eighth of the total address space of IPv6. However, the address space is so large that this is not considered a problem. Also, there is no separate prefix for anycast addresses; instead, they are ordinary global unicast addresses (with a few exceptions). Thus, while we can immediately differentiate link-local from site-local from global unicast from multicast addresses, there is nothing to differentiate anycast from global unicast addresses, except in how they are used.

A rich set of multicast addresses is defined in IPv6. One of particular interest to us here (partly because we will see it being used in neighbor discovery, in Section 9.3.3) is the concept of solicited-node multicast addresses. Any unicast or anycast address has a corresponding solicited-node multicast address, formed by appending the well-known 104-bit solicited-node multicast address prefix with the 24 lowest-order bits of a unicast or anycast address. Since interfaces on multiple nodes in theory could share the same 24 lowest-order bits, multiple unicast addresses map to each solicited-node multicast address. However, like a hash function, typically each unicast address on a link maps to a different solicited-node multicast address.

9.2.3   Coexistence of IPv4 and IPv6

As might be expected, the world will not switch from IPv4 to IPv6 overnight. There will be a period of technology transition, initially with pockets of IPv6 networks surrounded by a sea of the IPv4 Internet, then gradually a rise in the percentage of the Internet comprising IPv6 networks, and eventually the establishment of IPv6 as the dominant Internet protocol. Therefore, IPv4 and IPv6 need to coexist for a while. Basic approaches to interworking and coexistence of IPv4 and IPv6 include the following:

- Tunneling of IPv6 packets within IPv4 networks; in other words, IPv6 packets are encapsulated within IPv4 packets in IPv4 networks. One way to do the tunneling is to configure point-to-point tunnels for carrying IPv6 packets over IPv4 networks. However, it is also possible to tunnel IPv6 packets automatically over IPv4 networks if *IPv4 compatible IPv6 addresses* are used [4].
- Use of special NATs between IPv4 and IPv6 networks, where the NATs will handle the needed address translations.
- Isolated IPv6 hosts not connected to an IPv6 router, but if attached to a suitable IPv4 network, can use the IPv4 network as a virtual Ethernet, multicasting IPv6 packets to and from other isolated IPv6 hosts [5].

9.3   IPv6 Selected Procedures

In this section, we describe details of how certain key functions are performed in IPv6.

9.3.1   Header Processing

The IPv6 header contains a source and a destination address that are each four times larger than their IPv4 counterparts. One of the debates that occurred in the design of IPv6 was whether to keep using fixed-length addresses, as in IPv4, and only to increase the length of the addresses, or to use variable-length addresses. Variable-length addresses are arguably more scalable, but it was decided that the extra complexity was not worth the effort. Besides, using 128 bits provides for a very large address space, with many more addresses than needed for the foreseeable future.

The long address size was one reason why the rest of the header needed to be stripped to the bare essentials to avoid using excessively long headers. Many header features either do not need processing at every hop between source and destination or are not used in the great majority of cases. These are relegated to *extension headers*. Extension headers follow the IPv6 header and are typically arranged so they are easily processed in linear sequence (a router does not need to skip earlier headers to process later headers). To support optional features for which every hop in the path needs to be involved in the processing, a Hop-by-Hop Options header is defined (it may be used to support RSVP, for example). Otherwise, most of the extension headers do not need hop-by-hop processing, and they include:

- Fragment header—for use by the destination only, for reassembly of fragments;
- Routing header—for source routing; needs to be processed only by the nodes in the source route;
- Destination Options header—miscellaneous options meant for processing only by the destination (an exception is when source routing is used, as there might be another destination options header immediately before the Routing header, meant for processing by all the nodes in the Routing header);
- Authentication header—IPsec authentication header, for security;
- Encapsulating Security Payload header—also for IPsec.

The upper-layer header (e.g., TCP header) is placed last in the sequence of headers, and it is processed only at the destination. Thus the extension headers are sandwiched between the IPv6 header and the upper-layer header. We have seen that the IPv6 header has a Next Header field indicating the type of the extension header or upper-layer header immediately following the IPv6 header. Each extension header begins with a Next Header field that indicates the type of the next header. This makes it convenient for a router to know when it can safely ignore the rest of the headers.

In IPv4, every node between the source and destination must be prepared to perform fragmentation and reassembly operations. Since each link can have a different MTU, and no node in the path knows where the links with the shortest MTUs are, each node must be prepared to perform fragmentation. Similarly, each node must be prepared to perform reassembly. In IPv6, the design space has changed. Thus, it is too costly to have every hop be prepared to perform fragmentation and reassembly. Instead, fragmentation only happens at the source, and reassembly at the destination. This can be done because a minimum MTU size of 1,280 bytes is imposed in IPv6. Thus, packets smaller than this will definitely not need to be fragmented (not fragmented at the IP level, that is, for any link with an MTU less than 1,280 bytes; responsibility for fragmentation and reassembly *over that link* is relegated to the link layer). In order to take advantage of MTUs larger than 1,280 bytes, a path MTU discovery procedure is also available in which a source can discover the smallest MTU along a given path.

9.3.2 Address Autoconfiguration

One of the major benefits of IPv6 is the enhanced autoconfiguration, plug-and-play features it provides, which will relieve the burden on system administrators. There are two main types of address autoconfiguration: stateless and stateful. Stateless autoconfiguration is where there is no network server keeping a record of the address information of the nodes, while stateful autoconfiguration is where there is such a server, such as the DHCPv6 (DHCP for IPv6) server.

Stateless autoconfiguration generally occurs in two stages [6]. The first stage is where the node acquires a link-local address that can be used for communications on the link only. The second stage is where the node acquires other addresses, such as site-local and global addresses, for internetwork communications. In cases of an isolated network (e.g., a group of laptops setting up an ad hoc WLAN network at a conference), stage one is sufficient, and stage two is not completed. Also note that both hosts and routers follow the same procedure for autoconfiguring a link-local address on a link, so stage one is common to both hosts and routers, whereas stage two applies to hosts only (so the system administrator still has to configure the routers, and should therefore still have work to do!).

In the first stage, a node that first connects to a link can select a link-local address very conveniently, thanks to the huge address space available in IPv6. Since MAC addresses are usually unique and built into the hardware of network interface cards, the node quickly produces a candidate link-local address by concatenating the well-known 64-bit high-order bits reserved for link-local addresses with a 64-bit interface identifier. Today, MAC addresses are 48 bits long, so there is a procedure to pad these to 64 bits for use as an interface identifier that is very likely unique. In the future, MAC addresses may become 64 bits long, and then they can be used directly as interface identifiers.

We say that this address is only a candidate link-local address, because the node should make sure that no other node on the link is using it. The procedure for this check is known as duplicate address detection. This procedure is part of the neighbor discovery functions in IPv6 that will be covered in the next section. Briefly, though, the node sends out a *Neighbor Solicitation* message with the candidate link-local address. If it receives a response, a *Neighbor Advertisement* message from another node for that link-local address, the address is already in use. In such an unlikely event, address autoconfiguration cannot proceed, and the node must be manually configured. Otherwise, the node can safely assign the candidate link-local address to its interface on the link and be able to communicate with other nodes on the link.

The second stage is only carried out by hosts. The hosts need to figure out if any routers are on the link, in which case the link is part of a larger network. Discovering such information is part of neighbor discovery, which will be discussed in greater detail in the next section. If no routers are discovered, the host should try stateful autoconfiguration. If routers are present, they provide information to the host in their *router advertisements,* such as whether to use stateless or stateful autoconfiguration, and prefix information that hosts can use for generating site-local and global addresses. Once the relevant information is obtained from router advertisements, the host can complete the second stage by autoconfiguring site-local and global addresses. Since these addresses are formed by concatenation of the host's link identifier with various prefixes, it is not necessary to test again for uniqueness, as uniqueness has been tested in stage 1, with the link-local address using the same link identifier.

As for stateful address autoconfiguration, DHCPv6 has been recently standardized as one means of carrying it out [7]. To date, no other mechanism has been specified. DHCPv6 relies on the host to already have a link-local address before it interacts with DHCPv6 servers. Unlike in DHCP for IPv4, where the request message is broadcast, DHCPv6 uses a special multicast address, the all_DHCP_relay_agents_and_servers address, to which the host sends a request for stateful autoconfiguration, over UDP. In one of the modes of usage, a suitable DHCPv6 server responds with a DHCPv6 reply. It provides the host with the requested addresses, and also with other configuration information like the address of DNS servers. It is also perfectly legitimate for a node to obtain its addresses through stateless autoconfiguration, and still contact a DHCPv6 server to obtain other configuration information about DNS servers.

When would stateless autoconfiguration be preferred to stateful autoconfiguration, and vice versa? Stateless autoconfiguration is preferred when a site cares less about the exact addresses that hosts use, except that they are routable and unique, whereas the stateful approach is used when more control is desired. The stateless approach is more convenient in some ways, because DHCPv6 servers are not required. As a network size grows, it gradually makes more sense to use stateful autoconfiguration for more control over address management.

9.3.3   Neighbor Discovery

Neighbor discovery is a cornerstone of the IPv6 design [8]. Neighbor discovery is not so much a single function as a group of capabilities and functions allowing a node to discover information about a link to which it is attached, as well as information about neighbors on that link (i.e., other nodes attached to the same link). Neighbor discovery supports a number of major functions, including:

- Address autoconfiguration.
- *Router discovery:* IPv6 allows hosts to find routers on the link to which they are attached, using this procedure.
- *Next-hop determination:* This procedure is to decide the next-hop IP address to reach a particular destination; it may be the IP address of a router or the destination itself.
- *Discovery of other link information:* This procedure is to discover important information about the link, such as the link prefix and parameters like the link MTU.
- *Address resolution:* It is no longer necessary in IPv6 to have a separate protocol for address resolution, such as ARP in IPv4; instead, this capability is part of neighbor discovery.
- *Neighbor unreachability discovery:* This procedure is to find out which neighbors are no longer reachable.
- *Duplicate address detection:* This procedure is to find out if a candidate address (for use by a node) is already in use by another node.

Neighbor discovery makes use of *router advertisements* and *router solicitations,* and *neighbor advertisements* and *neighbor solicitations,* to accomplish much of the desired functionality.

IPv6 uses router advertisements from routers to provide the information needed for router discovery, next-hop determination, and discovery of other link information like the link prefix and link MTU. Router advertisements are only sent once every few minutes, to avoid bandwidth wastage. However, hosts may send out router solicitations (soliciting for router advertisements) in order to hear router advertisements sooner. The router discovery and next-hop determination procedures supersede the need for a default gateway, as used in IPv4.

Meanwhile, neighbor solicitations and neighbor advertisements are used for address resolution and duplicate address detection, and sometimes also for neighbor unreachability detection. The usages of these versatile messages is shown in Figure 9.4. Neighbor solicitations are sent by nodes that do not know the link-layer address associated with a particular unicast address (for address resolution), to detect if a particular unicast address is in use (for duplicate address detection), or to detect if a neighboring node is still reachable (for neighbor unreachability detection).

For address resolution, without knowledge of a link-layer address to use for the message, the sending node has to rely on the solicited-node multicast address of the target node. Since all interfaces must listen for the solicited-node multicast address corresponding to each associated unicast address it is using, the target node will receive the solicitation and can respond to the solicitation by unicasting a neighbor advertisement back, providing its link-layer address.

For duplicate address detection, what is of interest is whether or not anybody sends a neighbor advertisement in response to the solicitation. A response indicates a duplicate address, and the lack of a response indicates that the address is not a duplicate. In any case, we see how the neighbor solicitation/advertisement exchange is

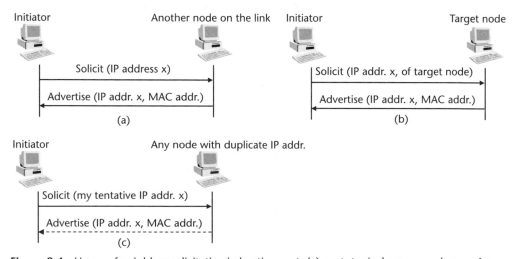

Figure 9.4 Usage of neighbor solicitation/advertisement: (a) prototypical usage; and usage for (b) address resolution and (c) duplicate address detection.

more efficient than the ARP used in IPv4. The solicitation is not broadcasted on the local subnet, but sent to the solicited node multicast address. In the unlikely event that there are multiple matches to that multicast address, the information in the neighbor advertisement should nevertheless still be sufficient to provide the correct link-layer address.

Sometimes, neighboring nodes become unreachable, perhaps because the neighboring node has failed. Neighbor unreachability discovery is used to detect such unreachability. It is only used in the context of nodes to which a node is actively sending packets. If upper-layer protocols can provide hints that the deliveries are being successfully made, the network layer is assured that the node is reachable. However, if such hints are unavailable, then the node sends a neighbor solicitation to the node in question, as a probe. If it receives a neighbor advertisement back from the node, all is well. Otherwise, it has detected that that neighbor is unreachable.

Neighbor discovery supports address autoconfiguration in a number of ways:

- Router advertisements provide information needed by address autoconfiguration. This includes the link prefix, and routers can also specify if hosts should use stateful or stateless address autoconfiguration.
- Duplicate address discovery is used during autoconfiguration to ensure that the autoconfigured address is unique.

9.4   IPv6 and Wireless

What are the implications of using IPv6 with wireless networks? Compared with IPv4, the enhanced autoconfiguration features of IPv6 are very useful for wireless networking. Stateless autoconfiguration, for example, may be very useful in cases where a group of mobile devices wish to communicate in a makeshift, temporary manner, perhaps at a temporary gathering like at a conference, a convention, or an event in a stadium. The autoconfiguration feature is one of the ways that IPv6 provides better mobility support than IPv4, as we will see in Section 9.4.1.

Unfortunately, as already mentioned, one price that we pay for the larger 128-bit address space in IPv6 is larger headers. Large headers have the most impact (in terms of high overhead) on short packets, such as those used for multimedia over IP (long packets should not be used because they add to the end-to-end latency, as discussed in Chapter 4). Schemes like RTP multiplexing and header compression become more crucial in this case.

9.4.1   Mobile IPv6

Mobile IPv6 (MIPv6) is similar to MIPv4, except that it works with IPv6 instead of IPv4 [9]. Its basic purpose is to enable a roaming mobile host to continue to receive packets originating from applications in correspondent hosts that address the packets to the home (permanent) IP address of the mobile host. IPv6 is more friendly towards mobile hosts than IPv4, because IPv6 was designed with the importance of mobility support in mind. Certain functionality is built-in to IPv6 that supports MIPv6 (thus, it is sometimes said that mobility support is integrated with IPv6). For example, MIPv6 enjoys advantages over MIPv4 such as the following:

- IPv6 includes more powerful address autoconfiguration capabilities. Therefore, each MH should be able to obtain a globally routable address in a foreign network, alleviating the need for FAs. In IPv4, a MH in many cases has no means of obtaining an address in a foreign network other than through a FA (unless, for instance, DHCP is used or it is connecting through GPRS).
- In MIPv6, route optimization is a standard feature, whereas it was a nonstandard option in MIPv4 and not widely supported. All IPv6 hosts are required to understand binding updates, unlike IPv4 hosts (however, a MH is not required to send binding updates to all CHs).
- When a CH sends packets directly to the COA of a MH, it does not need to encapsulate the packets, thus not incurring encapsulation overhead. Instead, a new routing header for source routing has been defined for this purpose in IPv6.
- Rather than using special messages, binding updates can be piggybacked on ordinary IP packets, through the use of the new mobility header (an IPv6 extension header). By piggybacking, we mean that ordinary IP packets are sent, but the mobility header is added to these packets to convey MIPv6-related information.

MIPv6 works as follows (see Figure 9.5): When an MH finds itself in a foreign network, it will autoconfigure an address in the foreign network. This address will be suitable for use as its MIPv6 COA. The MH will then register with its HA, by sending a binding update to the HA, which will return a binding acknowledgement. The exchange of binding update and binding acknowledgment must use IPsec ESP for security purposes. The MH may also send binding updates to some or all of the CHs with which it is communicating. Before sending such binding updates to CHs, it needs to execute the return routability procedure, for security purposes. The return routability procedure is illustrated in Figure 9.6 and will be discussed further in Section 9.4.2 in conjunction with security issues.

There are two ways that communications between an MH and a CH may occur. One way is when an MH does not inform the CH about its current COA. Packets from the CH go to the HA (the HA in MIPv6 is an enhanced IPv6 router in the home network), and are tunneled to the foreign network, as in IPv4. However, for the return path, instead of going direct to the CH, packets are reverse tunneled to the HA, using the COA as the outer source address and the MH's home address as the source address in the encapsulated packet. From the home network, the packets are forwarded to the CH. This solves a variety of firewall-related ingress filtering and egress filtering problems, at the cost of quadrilateral routing (instead of triangular routing).

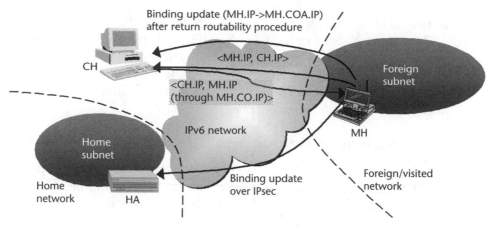

Figure 9.5 Overview of MIPv6.

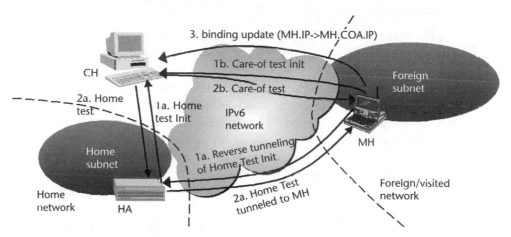

Figure 9.6 Illustration of return routability.

The other way an MH and a CH communicate is using route optimization; it is expected that most communications from CHs to MHs will use route optimization [9]. The MH sends binding updates to CHs whenever it moves to a new foreign network. After receiving a binding update, a CH can send packets to the MH using source routing, thus eliminating the encapsulation overhead found in MIPv4. As discussed in Section 9.3.1, source routing works in IPv6 using an IPv6 extension header known as a routing header. There can be different types of routing headers in IPv6, to allow for different rules and treatment of routing headers. The prototypical routing header for general source routing is type 0, whereas a new routing header is defined for MIPv6, and designated type 2. This allows firewalls to treat these packets differently from regular source routed packets, if so desired. Also, more stringent and specialized rules apply to type 2 routing headers. They can only contain one address, and the node that processes the routing header must verify that the address contained in the routing header is indeed its home address. In the other direction, for packets from the MH to the CH, the MH sends the packets with its COA as the source address, avoiding firewall-filtering problems that would be encountered if the home address were used. But how about the home address? It is still included, in the *home address option*. This is simply a type of IPv6 destination options extension header containing the home address.

9.4.2 Security Issues in MIPv6

In the early stages of MIPv6 development, the proposed concept of piggybacking of binding updates on data traffic packets was trumpeted as one of the highlights of MIPv6, as eliminating the need for separate packets to be sent just for binding updates. However, concerns soon arose regarding the need for security for the binding updates.

As we have seen in Chapter 7, in MIPv4 the MH and its HA must have a security association so that MIP registration can be authenticated. In MIPv6, IPsec ESP is used instead, providing more security (not just message authentication, as in MIPv4, but also data integrity). However, there are unresolved issues regarding the interaction of piggybacking and IPsec. In the case of binding updates sent to CHs for route optimization, the use of IPsec encounters the same issues, such as interaction of piggybacking and IPsec. However, there are other issues as well. Unlike the case of the HA, where it can be reasonably assumed that a security association exists between the MH and HA, it is more difficult to assume that a security association exists between the MH and every potential CH. The decision was therefore made not to require the same level of security for binding updates to CHs as to the HA (but the decision may be revised in the future, as MIPv6 continues to evolve); the result is the return routability procedure.

The return routability procedure is not a regular cryptographic protocol, but a procedure meant to limit the threat to a level of security not worse than in the IPv6 Internet. Basically, the idea is to test that both the home address and COA are routable to the MH (hence the name return routability), and if so, binding updates can be accepted. The way it works is that the MH initiates the return routability procedure by sending a *home test init* and a *care-of test init* message to the CH, with the source address being the home address and the COA, respectively (hence, the home test init is sent via the home agent after reverse tunneling to the home network). The CH responds by sending two tokens to the MH, one through its home address, in a *home test* packet, and the other through its COA (as obtained from the care-of test init message), in a *care-of test* packet. Only if the MH receives both binding updates can it then combine the two tokens to create a binding key. The binding key is the shared secret used for the MAC to protect the integrity of actual binding updates subsequently sent by the MH. The procedure is illustrated in Figure 9.6. The sequence of messages is numbered, where 1a and 1b are sent at the same time, and then 2a and 2b are sent in response, only after which the binding update (message 3) is sent.

9.5 Summary

In this chapter we are only able to discuss briefly the security issues related to MIPv6. MIPv6 and especially its security aspects are still works in progress as of the time this book is being written. More details on security aspects can be found in Nikander [10].

References

[1] Deering, S., and R. Hinden, "Internet Protocol, Version 6 (IPv6) Specification," RFC 2460, December 1998.

[2] Brown, S., et. al., *Configuring IPv6 for Cisco IOS*, Rockland, MA: Syngress Publishing, 2002.

[3] Hinden, R., and S. Deering, "Internet Protocol Version 6 (IPv6) Addressing Architecture," RFC 3513, April 2003.

[4] Gilligan, R., and E. Nordmark, "Transition Mechanisms for IPv6 Hosts and Routers," RFC 2893, August 2000.

[5] Carpenter, B., and C. Jung, "Transmission of IPv6 over IPv4 Domains Without Explicit Tunnels," RFC 2529, March 1999.

[6] Thomson, S., and T. Narten, "IPv6 Stateless Address Autoconfiguration," RFC 2462, December 1998.

[7] Droms, R., et al., "Dynamic Host Configuration Protocol for IPv6 (DHCPv6)," RFC 3315, July 2003.

[8] Narten, T., E. Nordmark, and W. Simpson, "Neighbor Discovery for IP Version 6 (IPv6)," RFC 2461, December 1998.

[9] Johnson, D., C. Perkins, and J. Arkko, "Mobility Support in IPv6," IETF draft-ietf-mobileip-ipv6-24.txt, June 2003, work in progress.

[10] Nikander, P., et al., "Mobile IP Version 6 Route Optimization Security Design Background," IETF draft-ietf-mip6-ro-sec-00, April 2004, work in progress.

CHAPTER 10
Services and Applications

Most consumers do not simply appreciate technology for technology's sake. They want to know what the technology can do for them. You give them a new gadget and they ask, "What is this device good for?" Our example family of the future from Chapter 1, Alice, Bob, Charles, and Diana, would probably represent this type of consumer. Indeed, although the focus of most of this book is the underlying networking technology that enables services and applications to be provided on wireless IP networks, it is the services and applications that will drive the commercial success of the technologies.

As explained in Chapter 1, the focus of this book is on the network layer aspects (and some link-layer aspects) of wireless Internet telecommunications. Thus, we only briefly discuss application development and execution environments for wireless IP services. Instead, we focus on going from the main coverage of the book—the network layer—to the closely related topic of middleware to abstract network capabilities and present them to application developers. In particular, among the exciting developments in such middleware is OSA, which is being developed by 3GPP and the Parlay Group, in Section 10.2. Implementation of these application program interfaces (APIs) is beyond our scope in this book, but I hope to convey the essence of the ideas even to readers who are not programmers. Since OSA is more focused on the network infrastructure part of the network, more terminal-centric technologies will be considered in Section 10.3.

10.1 IP Connectivity or More?

IP is a very flexible protocol, and the packet-switching and protocol-layering concepts of IP networks are very powerful. As a result, it gives engineers more headaches, in terms of choices that can be made, but it also gives more joy when these choices are well made. In this section, we discuss one of the choices that traditional wireless network operators face as they move towards greater use of IP in their wireless networks.

In recent years, designers of cellular networks such as UMTS have started embracing the concept of all-IP networks, recognizing the growing importance and dominance of IP. They began making IP play an increasingly bigger role in providing transport in the core network portion of cellular networks. In moving in this direction, one of the questions they faced was whether they would provide *IP connectivity only* or provide value-added IP-based services in addition to IP connectivity. In the case of providing IP connectivity only, the network operator sees itself as a provider of connectivity only, giving users the freedom to do whatever they like with that connectivity, such as obtaining services from various application service providers.

To obtain multimedia session management capabilities, the user would need to find the appropriate software to use from the appropriate source, and perhaps subscribe to SIP server service from dozens of potential vendors. With the flexibility of choices available, comes the responsibility to make good decisions. While this is the existing model for Internet users today, there might be many less technically sophisticated users who prefer to buy basic services in addition to IP connectivity, in a package from the network service provider.

In Chapter 12, we introduce the IP multimedia subsystem of UMTS, which can be viewed as one way to provide value-added IP-based services to users of the UMTS wireless network; the services provided would be multimedia session management services similar to voice session management services provided in systems like GSM, but the new services would be for multimedia over IP. This might appeal to customers who just want "a multimedia phone that works," and do not care about putting together their own package (which they could do if they were given just IP connectivity alone). We will come back to the IP multimedia subsystem and the value-added services it provides in Chapter 12, because we develop the discussion in the context of GSM evolution, and hence need Chapter 11 before we can fully address this issue.

A second customer for value-added services is not the mobile phone or mobile device user, but application service providers. The network operator has valuable information that can be useful for application service providers seeking to sell services to mobile users. For example, the network operator has each mobile user's location, which could be useful for location-based services. Also, the operator has each mobile user's subscription profile, nonconfidential parts of which could be useful for application service providers that may wish to provide user-centric advertising, for example. Furthermore, the network operator has certain capabilities (such as the ability to initiate and accept multimedia sessions from and to mobile users, such as provided by the IP multimedia subsystem) that can be useful for application service providers. Thus, if a network operator can provide access to these network capabilities, perhaps through some form of middleware that controls access and provides authorization and accounting features, the operator has grown from just providing IP connectivity to also providing some value-added services. We will see one example of such middleware in Section 10.2.

We note that no comparable choice existed in the traditional circuit-switched phone network. Since it used circuit-switching technology rather than packet switching, the traditional phone network required connection setup signaling before communications between two end points could commence. Switches would not switch any traffic otherwise. Thus, the ability to communicate over that network was integrated with the connection setup services in the intermediate switches. Everything was provided by the network operator. In contrast, for IP in cellular networks, the choice was available, and the momentum is currently in the direction of providing value-added services in addition to IP connectivity.

10.2 Middleware—Open Service Access

OSA is an application-enabling platform designed to facilitate the development of innovative new applications for telecommunications networks by third-party application developers [1]. It is desirable to bring in these third-party application developers because:

1. Network operators are earning declining revenues from merely providing transport capabilities. Operators need to add value to the bits transported over their network. One way to do this is to add services and applications over the network. The main problem with this is that the higher up in the protocol stack one goes, the harder it is to predict what services and applications will be financially viable. A solution is to open the network for third-party applications, where access is controlled by middleware—this is the OSA solution. The network operator thus controls the service-enabling platform. No matter which services and applications end up being successful, the network operator wins, because the network operator has a piece of the action, through its control of the middleware.
2. Third-party application providers, unlike the network operator, may specialize in application development. This, along with competition between multiple third-party providers, would foster rapid development and deployment of innovative applications.

Moreover, OSA is designed so that the pool of available third-party application developers will be large, and in principle could be the same as the general pool of information technology application developers. It does this by abstracting the network capabilities through APIs, so that the developers do not need to have specialized knowledge of wireless networks. Whereas the number of programmers with specialized knowledge of wireless networks may number in the thousands, the general pool of information technology application developers is in the millions. Thus, OSA enables this large pool of application developers to be tapped.

The OSA concept is also known as Parlay [2, 3]. OSA is the name given to the concept after 3GPP adopted ideas from the Parlay working group. The Parlay working group is a consortium founded in 1999 that is working on facilitating the convergence of communications and computing. It noted that the information technology industry has much experience and many tools for rapid creation of services and applications. Meanwhile, the environment for service creation in telecommunications was not as advanced; there were fewer experts, and specialized knowledge was needed to create telecommunications services. Thus, the Parlay working group promoted the notion that the capabilities of telecommunications networks could be abstracted through APIs, so that service creation for these networks was separated from the prerequisite of needing the specialized knowledge.

10.2.1 How OSA Works

OSA can be thought of as the next logical step in the evolution of application-enabling architectures. Before intelligent network (IN) concepts were introduced into telecommunication networks, services were built into the hardware of switches. If a new application (such as toll-free calling) were desired, the switch would need to be replaced with one that had the additional services. A major innovation of the IN concepts was the separation of the transport plane from the service plane. If the service plane was implemented in software, the same addition of a new application might merely require a software upgrade. The transport plane, with its switching fabric, would remain unchanged. Therefore, the implementation of IN concepts made it much more convenient to create new services, and to rapidly prototype, test, and build new applications that use these new services.

Notwithstanding the benefits of IN, application development still required a thorough understanding of protocols and infrastructure. Bugs could potentially bring down the whole network. Therefore, network operators were very careful in developing new services and applications using IN. Furthermore it was risky to let third-party providers run applications over their network. Moreover, for wireless networks, the desire to support IN services with roaming between operator networks meant a slow, detailed standards process to ensure proper interoperability.

Moving beyond IN, there is much recent interest in open APIs for middleware in telecommunications networks. OSA is one of the results. The underlying network capabilities are abstracted and accessed through APIs. These APIs are open and well defined, and thus simpler to access than vendor specifications. These APIs allow applications to become independent of underlying network technology and vendor-specific interfaces. The use of APIs also allows controlled, secure, and stable access to network capabilities without putting the whole network in jeopardy if there are bugs in the applications. OSA is a good example of layered design principles put into practice.

There are three main components in the OSA functional architecture (see Figure 10.1): (1) applications, (2) framework, and (3) service capability servers (SCSs) providing service capability features (SCFs). Three sets of interfaces are also specified, namely (1) the framework interfaces between applications and the framework, (2) the network interfaces between applications and the SCSs, and (3) the internal API between framework and SCSs. The framework controls OSA access by the applications, facilitates discovery of SCFs by applications, and also has administrative functions. In addition, SCSs can register SCFs with the framework using the internal API.

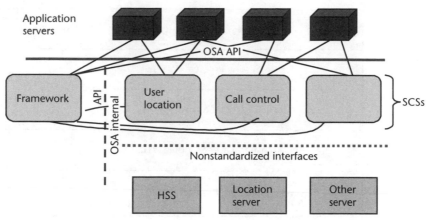

Application
servers

OSA API

Framework

API
OSA internal

User
location

Call control

SCSs

Nonstandardized interfaces

HSS

Location
server

Other
server

Figure 10.1 OSA.

From the network architecture perspective, there is an OSA gateway between the servers on which the applications reside, on one hand, and the service capability servers in the network, on the other hand.

10.3 Wireless Device Service–Enabling Technologies

We now briefly overview what we call wireless device service–enabling technology. The idea is that wireless devices are small, with limited memory, processing power, and display screen size, and the wireless link is typically of lower bandwidth, and with higher latency and higher error rates, than wired links. In the face of these constraints, wireless device service–enabling technologies enable wireless devices to provide services and applications like Web browsing. Given the huge popularity of Web browsing in the wired world, how can we allow mobile phone users to also browse the Web? It is reasonable to expect that Web browsing on these devices may be limited in some ways, but at least the essential capabilities should be feasible. The WAP forum (now part of the Open Mobile Alliance) was founded to develop a viable solution [4].

The result was the WAP-enabled phone, which supports a form of limited Web browsing with a WAP browser. In order to support the network communications of the WAP browser, the WAP forum defined a special protocol stack, optimized for the wireless device constraints and the wireless link characteristics. This can be seen on the left side of Figure 10.2 as the original WAP protocol stack. For example, wireless TLS (WTLS) is a lightweight version of TLS, as mentioned in Chapter 8. However, in the more recent WAP 2.0 specifications, WAP moved towards the use of regular protocols, albeit with wireless profiles. Thus, as seen on the right side of Figure 10.2, TCP is used, but with a wireless profile, and so it is called WP-TCP. Both the original and the new stack support the wireless application environment. The main application supported by the Wireless application environment so far has

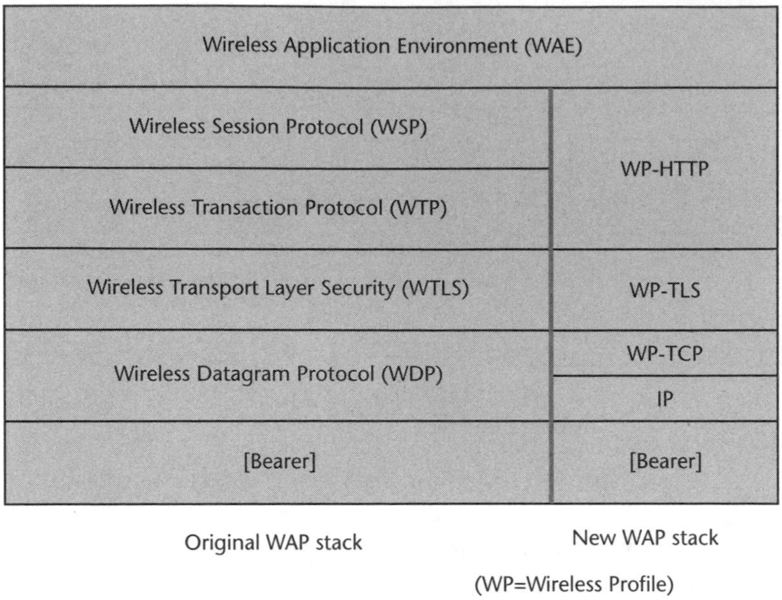

Wireless Application Environment (WAE)	
Wireless Session Protocol (WSP)	WP-HTTP
Wireless Transaction Protocol (WTP)	
Wireless Transport Layer Security (WTLS)	WP-TLS
Wireless Datagram Protocol (WDP)	WP-TCP
	IP
[Bearer]	[Bearer]

Original WAP stack   New WAP stack

(WP=Wireless Profile)

Figure 10.2   WAP protocol stack.

been Web browsing, for which a customized markup language, Wireless Markup Language (WML), has been specified. WAP uses WML instead of Hypertext Markup Language (HTML).

How does Web browsing with WAP work? There are many Web servers in the world today, and they understand HTML. One approach to providing contents for WAP-enabled phones is to require that as many of these Web servers as possible add support for WML in addition to HTML. The larger the number of Web servers that add WML support, the more useful WML becomes, and the more successful WAP is. However, this approach runs into the technology transition problem. Thus, the chosen solution was to introduce the concept of WAP gateways that would be placed in the path between WAP-enabled phones and Web servers, and that would do the translation between HTML and WML. Thus, immediately, WAP-enabled phones were able to browse a large selection of translated Web pages even though the Web sites only provided HTML content. Another benefit of WAP gateways is that they can also act as SSL/WTLS gateways, using SSL for the communications segment between the gateway and the Web server and WTLS for the communications segment between the WAP-enabled phone and the gateway. Moreover, there need not be a one-to-one relationship between Web servers and WAP gateways—a typical WAP gateway serves as a general gateway to the World Wide Web of Web servers. Figure 10.3 illustrates these ideas.

One of the drawbacks of WAP is the need for WAP gateways (or for Web servers to add support for WML in addition to HTML). Is there another way, one that allows Web servers to communicate directly with mobile phones without the need to support another markup language and without the need for servers, while at the

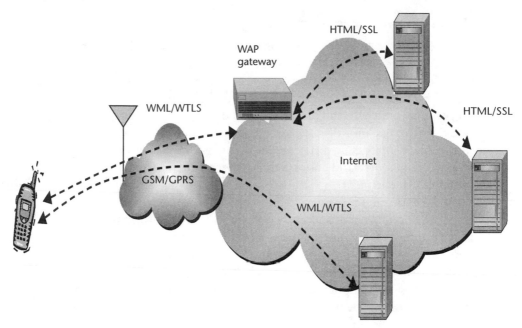

Figure 10.3 WAP network architecture.

same time taking into account the limitations of mobile devices? There is such a solution, and it is used in i-mode. Rather than introducing a new markup language like WML, i-mode uses compact HTML (cMTML), which is roughly a subset of HTML, without certain features like frames and tables.

10.4 Applications of the Future

For each new device or platform, the killer applications that emerge depend strongly on the state of the technology contained in the device. For example, in the early days of 3G system deployment, video telephony emerged as one of the most popular, if not the most popular, application for the new systems. Compared with 2G systems, the 3G systems support higher data rates, and thus can support decent quality video telephony (video telephony on the 2G systems would have to use lower-quality video and therefore be much less attractive to consumers). The wireless devices are also changing, with the lines between communications device and computing device blurring. For example, recent *smart phones* combine the features of mobile phones and personal digital assistants (PDAs).

What will applications of the future look like? Making predictions is a challenging task, since it depends on what the devices and platforms of the future can support. Nevertheless, it appears reasonable to speculate that some applications of the future will incorporate emerging concepts like location-based services and presence.

10.4.1 Location-Based Services

The concept of location-based services encompasses a broad spectrum of services with one thing in common: in one form or another, information on the location of a mobile device makes a difference in the service provided. For example, a location-based pizza store locator service connects the mobile phone to different pizza stores depending on where the phone is located. Location-based advertising services provide a different mix of advertisements depending on the device location, such as more male-interest advertisements if the phone is in the men's department of a department store, and more female-interest advertisement if the phone is in the women's department.

Since location information is needed in order to provide location-based services, the location of devices and users becomes valuable information. Network operators have control of this information, since they control the network and the elements and processes for obtaining this information. Thus, network operators can sell location information to application service providers. As we have seen in our discussion on OSA, one of the purposes of the mobility SCF in OSA is to provide user location information. So, one way that user location information could be sold to service providers would be through packaging with OSA.

10.4.2 Presence

Many people have used Internet messaging services, such as Yahoo messenger or AOL messenger or MSN messenger. The user can define a set of other users as buddies, whose online presence can be indicated in the application window. This is a good example of the concept of presence, which is about the dynamic status of the user, and can include more than just online status. For example, concepts of rich presence have been proposed that include more fine-grained information such as terminal capabilities. Some have even suggested that user location should be included as part of the concept of presence.

10.5   Summary

As wireless and IP have been converging, link, network, and transport protocols are being developed to handle issues such as QoS, security, and mobility. Meanwhile, it must not be forgotten that these are merely supporting components that allow services to be created and delivered to subscribers. The success of the applications and services in wireless IP networks will ultimately determine the success of wireless IP networks.

Therefore, in this chapter we offer a glimpse of some of the service-enabling tools for development of services for wireless IP networks. These include middleware concepts like OSA for bridging the gap between network capabilities and applications, and wireless device service enabling technologies that allow services (including Web browsing on small screens) to be supported on limited devices. We also briefly introduce concepts like location-based services and presence, which may be important parts of future applications for wireless IP networks. The coverage in this chapter is brief, and by no means comprehensive, as the focus of the book is more on the networking aspects of wireless IP networks. However, I hope it helps to convey some of the main ideas and encourages you to refer to other sources for more details.

References

[1]   3GPP TS 29.198-1 V 4.3.4, "Open Service Access (OSA); Application Programming Interface (API); Part 1: Overview (Release 4)," December 2003.

[2]   Glitho, R., and K. Sylla, "Developing Applications for Internet Telephony: A Case Study on the Use of Parlay Call Control APIs in SIP Networks," *IEEE Network Magazine,* May/June 2004.

[3]   Moerdijk, A.-J., and L. Klostermann, "Opening the Networks with Parlay/OSA: Standards and Aspects Behind the APIs," *IEEE Network Magazine,* No. 3, May 2003, pp. 58–64.

[4]   http://www.openmobilealliance.org.

CHAPTER 11
Evolution from GSM to UMTS

The popular GSM system is deployed all over the world in more than 197 countries, having provided cellular mobile services to over 863 million people (as of May 2003), and was estimated to have 1 billion people by 2004 [1, 2]. The primary service provided by GSM is mobile voice communications. We have briefly discussed GSM in Chapter 3. GSM, a 2G system, is evolving to UMTS, a 3G system (the high-level time line for this evolution is shown in Figure 11.1). UMTS is being designed to be a multiservice network that supports higher data rates and multiple new and innovative services (including a variety of data services, not just voice). Of the new wireless IP systems being developed around the world, it is arguably the one that is furthest along in its development and with the most momentum. Thus, UMTS has the potential to be the most widely deployed wireless IP telecommunications system in the world, within a decade.

In this chapter and Chapter 12, we discuss the evolution of GSM to UMTS to provide a perspective on the standards development process of a canonical wireless IP system. Along the way, various relevant issues will be discussed. This chapter is more on the evolutionary process, whereas the next chapter is focused on the IP multimedia sub-system that is foundational in providing IP multimedia services in UMTS. In these two chapters, then, we will see how the pieces fit together in a real wireless IP system.

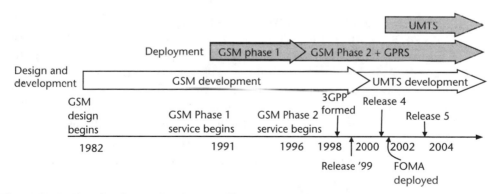

Figure 11.1 Time line for GSM and UMTS development and deployment.

11.1 From GSM to GPRS

11.1.1 Overview

GSM is not optimized for data traffic. In some ways, this is analogous to how the normal telephone system (with basic voice facilities, excluding ISDN or DSL) is not optimized for data traffic. Voice-band modems used over telephone lines have been tweaked over the years so that they can reach 56 Kbps these days, where they are hitting fundamental capacity limits. Although it is possible to use GSM for data traffic, the data rates are low because it is constrained to reuse the GSM time slot for data in place of voice. On the network side, the traffic is carried on voice trunks and circuit-switching equipment (like the MSC) that are optimized for voice, not data. It is expensive, again because it is constrained to reuse the GSM time slot for data in place of voice. Thus, valuable radio resources are not well utilized, and the subscriber is charged for the entire time the circuit (the time slot) is reserved for the subscriber, even if the traffic is bursty and there are stretches of one or more frames when the time slot is empty. From the user perspective, GSM data may take too long to connect; the alternative is that it is always on, which would be a very poor utilization of radio resources and much too expensive to be practical.

GPRS [3] is an enhanced packet service added to GSM systems. GPRS enhances GSM in a way loosely analogous to how digital subscriber line (DSL) enhances a regular phone line so it can provide high-rate data services—in addition to some changes beneath the network layer, additional network elements are provided to support the new services. The analogy is loose in that there are many differences between how DSL and GPRS provide their respective enhancements.

GPRS provides higher data rates by allowing the aggregation of two or more time slots to be used for the traffic to and from one MS (we tend to think of an MH as an Internet host, whereas we use MS to refer to something like a GPRS terminal that can be an Internet host, but is also more than that). Over the air, GPRS utilizes the radio resources more efficiently by dynamically allocating the time slots for different MSs and maintaining states related to the traffic activity of each MS (the GPRS mobility management states that we will discuss shortly). Having the GPRS mobility management states also allows shorter access times on the average, because terminals can be in standby state ready to quickly switch to ready state when there are packets to communicate. In the network, on the other hand, GPRS uses network resources more efficiently by adding a set of network elements that are parallel to the circuit-switching network elements like the MSC. Packet data traffic gets routed to and from these network elements, the GPRS support nodes (GSNs), where they are packet-switched. We thus see that both over the air, and in the network, statistical multiplexing is exploited for more efficient data traffic transport. Furthermore, GPRS provides simplified, direct access to packet data networks through the GSNs. The GPRS network architecture (simplified) is illustrated in Figure 11.2.

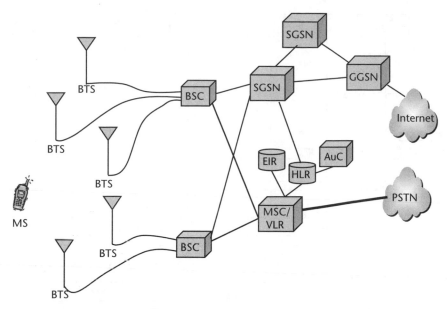

Figure 11.2 GPRS network architecture.

11.1.2 More Details

GPRS, as its name suggests, is not just for one kind of packet data, such as IP data. Rather, it is a generalized packet radio service that can support a variety of data traffic. Originally, GPRS support of IP and X.25 data was defined, but the references to X.25 support were removed in the year 2000. Nevertheless, GPRS in principle can support non-IP data protocols, whether X.25 or otherwise. In order to use GPRS services, an MS must be GPRS-capable. The network makes GPRS services available to a given MS after it successfully performs a *GPRS attach* procedure. The GPRS services are available until a corresponding *GPRS detach* procedure is performed. The GPRS attach and GPRS detach procedures are similar to the IMSI attach and IMSI detach used in GSM that were discussed in Chapter 3. A GPRS attach may be performed some time after an IMSI Attach, for instance, if a subscriber turns on a mobile phone to use GSM services, and only later wishes to use GPRS services. Alternatively, a joint attachment procedure could be used that would result in the MS becoming both IMSI-attached and GPRS-attached. This is illustrated in Figure 11.3.

Because multiple data protocols can be supported, GPRS defines the concept of *Packet Data Protocol (PDP) context*. Each PDP context is associated with one and only one data protocol, such as IP, and includes the related parameters like IP address and QoS requirements to support an instance (in the MS) of communications with that data protocol. Each of the data protocols, such as IP and X.25, is a different PDP type. An MS could simultaneously have multiple PDP contexts active, each associated with its own data protocol that in general may be the same protocol

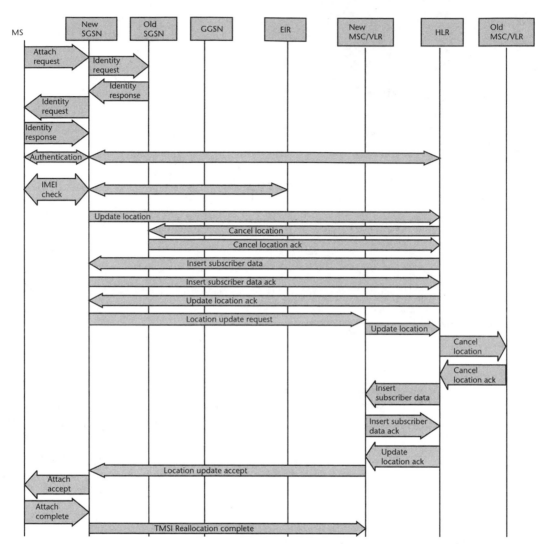

Figure 11.3 Combined GPRS/IMSI attach.

(but different address) as some of the other PDP contexts and different from some of the others. For example, an MS may have one or more contexts for IP data only, for X.25 only, or for both IP and X.25. When there are multiple PDP contexts of the same type, such as IP, each is treated by the external network (e.g., the Internet) as a different logical IP host with its own IP address and other parameters. Technically, protocol data units (PDUs) for a particular PDP are referred to as PDP PDUs, although we may also informally refer to them as user data packets. Before PDP PDUs may be carried by the GPRS network for a particular PDP type, a suitable PDP context must be activated. We will discuss the PDP context activation procedure later, after first introducing a few prerequisite concepts, such as functions of the GPRS network elements.

GPRS is designed as an extension to GSM networks, not as a new stand-alone type of network. Thus, over the air it reuses the GSM channelization scheme (e.g., frames with eight time slots). Thus, the same GSM base stations can be used to simultaneously support both GSM and GPRS services on different channels. However, a given time slot can only be allocated to provide GSM or GPRS services at any given time. The time-slot allocation is flexible in that one to all eight time slots in a frame can be allocated to GPRS services, and this number can change dynamically on a frame-by-frame basis, based on traffic loads and other considerations. The allocations for uplink and downlink are done separately, allowing asymmetric traffic loading to be handled efficiently. Unlike time slots allocated to speech, where each call has its own time-slot allocation, time slots that are allocated to GPRS are shared between active GPRS users. The configuration of GPRS channels is broadcasted to MSs on common control channels. GPRS also has features to monitor the dynamic usage of GPRS time slots for congestion or underutilization, and congestion control procedures to deal with congestion.

In the core network, a set of new packet-switching network elements is introduced, parallel to the circuit-switching core network elements (e.g., the MSCs). These network elements are called GSNs. In a GSM network that is upgraded to provide GPRS services as well, the BSs and base station controllers (BSCs) are shared by GSM and GPRS traffic. GSM and GPRS traffic take different paths between the BSC and the correspondent host. GSM traffic continues to be routed through MSCs, whereas GPRS traffic is routed through GSNs. The GSNs form a network subsystem connected over an internal IP-based network. This internal IP-based network may have its own private IP address space, and is limited in scope to only the GPRS network subsystem. It serves mainly a transport function—user data packets are tunneled between GSNs over this internal IP-based network (we will return to the tunneling concept), whether they use IP, X.25, or some other data protocol. That IP is one of the user data protocols supported by GPRS is incidental to the use of IP within the GPRS network. We will return to this point in Section 11.1.6.

There are two kinds of GSNs: serving GSNs (SGSNs) and gateway GSNs (GGSNs). Each SGSN is responsible for keeping track of the location of individual MSs within its service area (comprising many BSs and BSCs). It also performs security functions and access control (more details can be found in Section 11.1.5). Thus, the SGSNs are analogous to serving MSCs. Meanwhile, the GGSNs are the interfaces to external packet data networks, such as the Internet, or X.25 networks. Thus, each GGSN should be the entry point for packets of a particular PDP type that are routed from an external network of that PDP type through the GPRS network to the MS. Therefore, the address associated with a PDP context should be routable from the external network to the appropriate GGSN, for instance, for IP data, and the IP address used by the MS should be routable to the appropriate GGSN.

The user data is tunneled between the MS and the GGSN, in a manner that is not apparent to the user except that the time it takes for the packet to traverse the

GPRS network and get from the MS to the external data network and vice versa. Note that this does not mean the PDP PDU (the user data packet) is completely untouched over the entire path between the MS and the GGSN. In fact, functions like IP header compression do modify the packet, but in a completely reversible way so no modifications are apparent to the user. Thus the user experiences the GPRS service as a wireless access service through the GGSN to the access router in the external IP network.

In order to set up the tunneling to link the MS with the appropriate external network, a suitable PDP context must be activated. This is handled in a procedure known as *PDP context activation*. An MS needs to be GPRS attached before it can proceed with PDP context activation. The PDP context activation procedure is shown in Figure 11.4. In the event that IPv6 is used, then the procedure is modified by adding an IPv6 router advertisement (and optional router solicitation) after the PDP context activation, for address autoconfiguration. The GGSN, as the first-hop router, plays the part of the IPv6 router on the same link as the MS. The modified procedure is shown in Figure 11.5 (we omit steps 2, 3, and 5 of Figure 11.4 for conciseness).

User data is transferred transparently between the MS and the external data networks with a method known as encapsulation and tunneling: data packets are equipped with PS-specific protocol information and transferred between the MS and the GGSN. This transparent transfer method reduces the requirement for the PLMN to interpret external data protocols, and it enables easy introduction of additional interworking protocols in the future.

11.1.3 GPRS and Mobility

A GPRS terminal can be in one of three mobility management states at any given time. Similarly, the network also maintains a parallel set of states for the MS, as

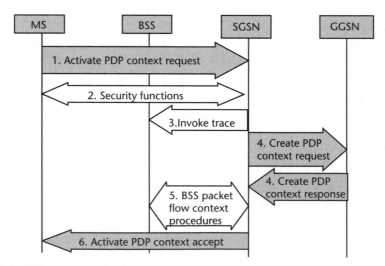

Figure 11.4 PDP context activation.

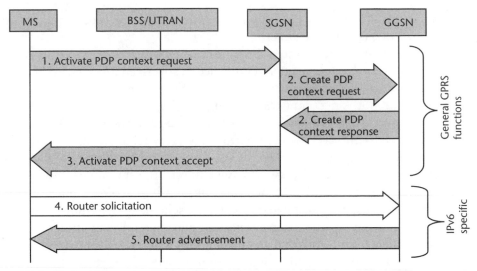

Figure 11.5   IPv6 address autoconfiguration.

shown in Figure 11.6. The three states are the idle state, the ready state, and the standby state. The idle state is when the terminal behaves as if it is "off" with regards to GPRS—no data can be transmitted to and from the MS in this state, and there is no GPRS location or routing information for the MS. We must distinguish between this GPRS idle state and the MS being really turned off. While in the GPRS idle state, the MS may be actively involved in GSM communications, and even if not, it continues to perform cell selection and do location updates (as is normal for

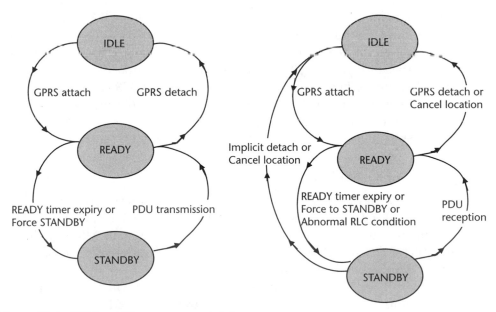

Figure 11.6   GPRS mobility management states.

an idle GSM phone, as we saw in Chapter 3). On the other hand, when it is really turned off, it is unreachable.

The ready state and standby state can be thought of as the states in which hand-off and location management take place, respectively. As we earlier saw in Chapter 6, one of the challenges in mobility management is how to tell when a data session is active or not, since data traffic, unlike voice traffic, is bursty. For example, for bursty traffic where the intervals between the gaps are not too large, you would not want to have the overhead and extra delay of switching from the standby state to the ready state each time the next burst comes. However, if the bursts are indeed large enough, it may indeed be worth switching to the standby state in between bursts, and back to the ready state each time the next burst comes. One might imagine that GPRS may use a sophisticated monitoring scheme to determine when a data session is active. Actually, the scheme is very simple—a ready state timer is used, where the MS "drops" from the ready to the standby state whenever the timer expires. The timer is reset whenever data is sent from the MS. Thus, the timer expiration implies lack of data traffic from the MS for the duration of the timer countdown.

Like in GSM, the MS is tracked to the cell level when it is actively communicating. For GPRS, this is when the MS is in ready state. The standby state meanwhile is a tradeoff designed to conserve radio resources and save power, while at the same time allowing communications to resume more easily than if the MS had returned to the idle state. Either the MS or network may initiate a GPRS detach procedure to move to the idle state. While in standby state, the MS updates the network with its location when it moves between routing areas (RAs), rather than between cells. RAs are like location areas for regular GSM idle mode, where both RAs and location areas comprise multiple cells. Thus, movement between RAs (or between location areas) is less frequent than movement between cells. Additionally, the MS periodically updates the network on its RA, even when it has been remaining in an RA for a long time. Therefore, if the MS is out of coverage and so does not send this periodic update, the network will be able to save resources by not paging the MS. At this point, the network (more specifically, the SGSN involved) has the option of performing an *implicit detach* (compared with the GPRS detach, where both MS and SGSN are involved, an implicit detach happens in the SGSN without the involvement of the MS). This sets the mobility management context for this MS to idle, in the SGSN, and it can then remove this context as well as the PDP context for the MS, and thus save resources.

11.1.4 GPRS and QoS

Four traffic classes are defined for GPRS QoS. These are conversational, streaming, interactive, and background classes, and we have seen them in Chapter 7 (Table 7.1) [4]. In order to support different QoS profiles on the same MS (e.g., different levels of QoS for different flows, perhaps each flow with different kinds of traffic and different QoS requirements), GPRS defines the concept of primary and secondary PDP contexts. If an MS uses only one PDP context, it will be a primary PDP context. Additional secondary PDP contexts can be activated that are associated with a primary PDP context, sharing the same IP address and other parameters, except for the QoS parameters. Each secondary PDP context may have its own set of QoS parameters. Figure 11.7 illustrates the concept of primary and secondary PDP contexts as a bundle of pipes. In PDP context activation, the MS requests that a particular set of QoS parameters be used. However, the SGSN may restrict the QoS parameters for the PDP context, based on SGSN-related factors like SGSN load and the MS subscription profile. The SGSN then passes on to the GGSN its recommended set of QoS parameters for the MS. This is referred to as QoS negotiated. The MS can either accept it or deactivate the PDP context.

How does GPRS treat traffic in different traffic classes differently? Each traffic class can be mapped to ranges of different QoS parameters, such as transfer delay (latency), packet error ratio, and whether delivery order is important. A variety of QoS mechanisms can then be applied to realize the desired QoS. For example, the link layer of the GPRS protocol stack, the RLC and MAC, supports four *radio priority levels,* plus another priority level for signaling. Header compression may be applied differently for traffic in different traffic classes. Traffic in different traffic classes may be assigned to different queues in the SGSN. The exact assignments, scheduling, and other functions are not specified, but are implementation-dependent. While also not specifying all the lower-layer QoS mechanisms, the layered UMTS QoS model that has evolved from GPRS is more structured and powerful, as we will see in Section 11.3.1.

At any time, a subscriber's HLR may send updated subscriber information to the SGSN, including possibly changes in the subscribed QoS profile. If necessary, the SGSN may then initiate a PDP context modification procedure to change the negotiated QoS.

11.1.5   GPRS and Security

GPRS builds on GSM security, adding additional security features specific to GPRS features (the security services it provides for packet-domain usage are basically equivalent to the services GSM provides for circuit-switched services). Thus, the MSC continues to play a central role, storing authentication triplets and being involved in the signaling for IMSI attach and location update procedures. However, it is the SGSN that plays this type of role for GPRS procedures like GPRS attach, RA updates, and PDP context activation. We consider the example of security functions in GPRS attach (Figure 11.3) and PDP context activation (Figure 11.4).

During the attachment procedure, we see in Figure 11.3 that step 4 is authentication. This is the same authentication algorithm used in GSM (Chapter 8), where

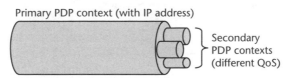

Figure 11.7   Primary and secondary PDP contexts.

the MS needs to return the correct SRES given a random number. However, in GPRS, since it is the SGSN doing the signaling with the MS, the SGSN (rather than the MSC) first needs to obtain the authentication information (triplets) from the HLR. The authentication signaling is shown in Figure 11.8. The authentication may also be done in step 2 of PDP context activation (Figure 11.4), detailed as "security functions."

11.1.6 GPRS and Wireless IP

Where and how does GPRS fit into the wider scheme of things, considering the context of wireless IP technologies in general? It helps to view GPRS as a dual-concept technology, meaning one that fits into the wider scheme of things in two different ways. First, GPRS can be viewed as an access technology, like WLAN, that provides IP connectivity to an IP-based network through an access router (the GGSN, in this case). Second, it is just as valid to view GPRS as a full-scale wireless IP network in its own right. The curious dual-view nature of GPRS is a direct consequence of the decisions of the architects of GPRS to include IP as one of the supported packet-data protocols while at the same time using IP as the basis of the internal packet-switching network in GPRS.

11.2   Moving Towards 3G

The push towards the 3G of mobile wireless systems has been driven by a number of developments, including the following:

* *The growth of multimedia applications that demand more bandwidth than traditional voice applications.* The 2G systems like GSM were designed to support voice as the main application. As we have already mentioned several times in this book, wireless IP telecommunication systems will need to

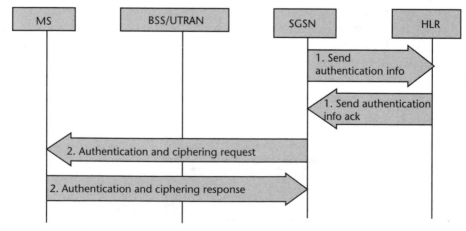

Figure 11.8 GPRS authentication signaling.

support multimedia applications, not just voice. Some of these applications need higher data rates than is necessary for telephone-quality voice transmission.

- *Competitive alternative to wired terminal access is desirable.* Wireless terminals have traditionally been a step or two behind wired terminals, in terms of features and capabilities. Wired terminal access provides higher data rates and, higher and more flexible quality of service, at lower costs than wireless systems. However, the wireless segment of the telecommunications market has grown rapidly, because of the great convenience of untethered communications with mobility support. Still, if wireless starts to lag too far behind wired, consumers' perspectives may change. The challenge to wireless terminal access is to provide comparable (to wired access) data rates and quality of service at competitive prices, in other words, without charging too much of a premium for mobility. With DSL and cable modem becoming more widespread, people are getting used to multimedia over IP and expect wireless to keep up. In short, wireless can remain healthy if it keeps up with wired, albeit a step or two behind, but may lose out to wired in the long term if it gets ten steps behind. Thus, wireless needs to evolve to keep "in step" and remain competitive and attractive.

- *In some cases, the 2G systems are near full capacity.* Either a redesigned 2G system or a new 3G system is the answer to provide the supply to meet the growing demand. It made more sense to go for 3G, given that this driver for 3G is only one of several. One of the reasons the first UMTS network began service in Japan (Freedom of Mobile Access, FOMA, by NTT DoCoMo) was the urgency felt there by their running out of capacity for voice traffic.

- *There is a powerful trend towards convergence of different systems and wireless access technologies.* More efficient and cost-effective service can be provided if there are fewer competing standards providing similar services. The first and second generations of cellular systems have seen many different systems. Even the dominant 2G system, GSM, only has about 70% of market share. This made it hard to roam globally with one's phone (more on the benefits of global roaming in the next bullet point). One of the original goals of the ITU for 3G systems was that there would be one global standard used everywhere. If it had succeeded, it would have eased the provisioning of global roaming. Other benefits include economies of scale in manufacturing.

- *Global roaming.* Global roaming allows the user to have wireless access (preferably with most of the user's desired features) anywhere in the world, and not just at the user's home location.

The motivations listed here, among others, drove the following requirements:

- *High-rate wireless access.* To meet the growing demand for wireless services, which increasingly demand more bandwidth, high-rate wireless access capabilities were required. High-rate wireless access was seen as especially important in the indoor and low-speed environments to enable wireless to be competitive with wired alternatives. (But this was before the boom in popularity of WLAN—today, many experts see cellular and WLAN systems as complementary, as we will see in Chapter 13, and the perceived importance of high-rate indoor access for 3G wireless has declined since WLAN is better suited for that arena.)

- *Multirate wireless access radio transmission technologies with flexible service requirements (quality, symmetry, and delay).* In order to serve a variety of different multimedia applications with varying demands on bandwidth, quality of service, link symmetry, and delay tolerance, flexibility is essential. This also helps in integration of different services (e.g., voice and data) into a single device. Good solutions to the problems of providing cost-effective high quality of service over radio links are necessary to enable wireless to be competitive with wired alternatives.

- *Small, lightweight, and convenient mobile terminals.* The main motivation for this requirement is to be competitive with wired alternatives. The availability of small, lightweight, and convenient mobile terminals will stimulate demand in addition to the already high projected demand, and it will be one of the factors allowing for economies of scale. Economies of scale will make wireless more competitive.

- *Global standardization.* This should allow for global roaming.

Additional requirements include the following:

- Some backward compatibility with and accommodation of existing systems. This would be helpful for the transition period.
- Smooth migration paths to the new systems. This is helpful to ensure acceptance and success of 3G systems.

Two main families of 3G systems are being developed. One is the GSM-evolved UMTS that uses the WCDMA air interface. The other is cdma2000, which is evolved from Qualcomm's IS-95 CDMA system. The aim of the ITU to achieve just one global mobile wireless system has therefore failed. Interestingly, there had originally been 15 proposals submitted to the ITU in 1998, and these were eventually consolidated into just these two systems plus a few other radio interfaces with marginal market presence. There had been efforts to harmonize the standardization of UMTS and cdma2000, but these ultimately failed to bring the two together, for both political and technical reasons. From the technical perspective, both systems can support high data rates and multiple data rates with flexible service support. The related phones and terminals are not bulkier than 2G phones.

However, the backward compatibility issues, and the strong desire of operators for a 3G system that can be reached smoothly through evolution of their existing 2G systems, were serious challenges that prevented a single global standard from emerging. Politically, it was very difficult, because there was one strong group that insisted on backward compatibility with GSM, whereas there was another strong group that insisted on backward compatibility with IS-95 CDMA. The former group coalesced into the 3G Partnership Project (3GPP) umbrella of standards organizations working on UMTS, whereas the latter group coalesced into the 3GPP2 umbrella of standards organizations working on cdma2000.

We discuss further the similarities and differences between UMTS and cdma2000 in Section 11.3.4.

11.3 UMTS

Release 1999 is the first of what might be called the true 3G mobile systems to be specified by 3GPP. It includes a completely new air interface, wideband CDMA (WCDMA), which was designed to use bandwidths of 5 MHz or more, to support the higher and variable data rates characteristic of 3G systems [5]. WCDMA was designed from scratch specifically for UMTS, and it is based on CDMA for the multiple access technology, rather than TDMA as used in GSM.

There are also changes in the network to support WCDMA and other features introduced with UMTS, and the network architecture (simplified) is shown in Figure 11.9. As in GSM, the portion of the network between the MSCs and the MSs is still arranged hierarchically, but the names of network elements have changed to reflect the differences from the corresponding GSM network elements. For example, the radio network controller (RNC) in UMTS corresponds to the BSC in GSM. Unlike the BSC, the RNC needs to be able to support soft handoffs (we introduced the soft handoff concept in Chapter 3). Thus there is a direct connection between neighboring RNCs to support soft handoff involving node Bs controlled by different RNCs. The MS is now known as the user equipment (UE), and the corresponding SIM it contains is the USIM (UMTS SIM). The core network elements like the SGSN and GGSN (for GPRS) and MSC retain their names, but the prefix "3G" is applied when there is a need to differentiate them from their GSM counterparts. A summary of correspondence in naming is shown in Table 11.1.

11.3.1 QoS

UMTS uses a layered QoS architecture, as shown in Figure 11.10 [4]. This is in keeping with good modular design principles, where each layer provides QoS-related services to the layer above so the topmost layer can provide end-to-end QoS services using the underlying network capabilities in a systematic manner. Within the UMTS network, the end-to-end QoS service layer makes use of the UMTS bearer service, which is an abstraction built over the radio access bearer service in the UTRAN and the core network (CN) bearer service in the core network. The UMTS bearer service attributes are shown in Table 11.2.

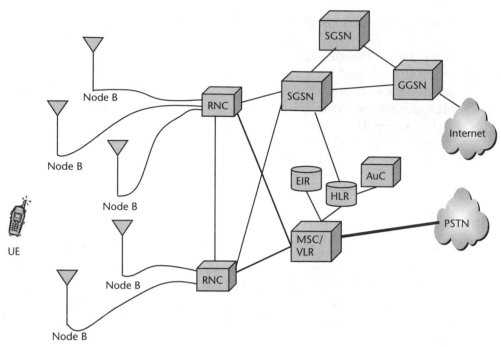

Figure 11.9 UMTS network architecture.

Table 11.1 Correspondences Between GSM and UMTS Network Elements

GSM	UMTS	Comments
Base Subsystem (BSS)	Radio Network Subsystem (RNS)	
Base Transceiver System (BTS)	Node B	Base station
Base Station Controller (BSC)	Radio Network Controller (RNC)	
Radio Access Network (RAN)	UMTS Terrestrial Radio Access Network (UTRAN)	
Mobile Station (MS)	User Equipment (UE)	
Subscriber Identity Module (SIM)	UMTS Subscriber Identity Module (USIM)	
SGSN	3G-SGSN	"3G" prefix can be dropped if
GGSN	3G-GGSN	context is clear
MSC	3G-MSC	

The control architecture for QoS management for the UMTS bearer service is shown in Figure 11.11. It shows that translation functions are needed to convert between signaling for external QoS protocols and the internal service primitives for UMTS bearer service. There are service manager functions in the mobile terminal (MT), the UTRAN, and at both edges of the CN, in other words, the border with the UTRAN and the border with the external networks. These service manager

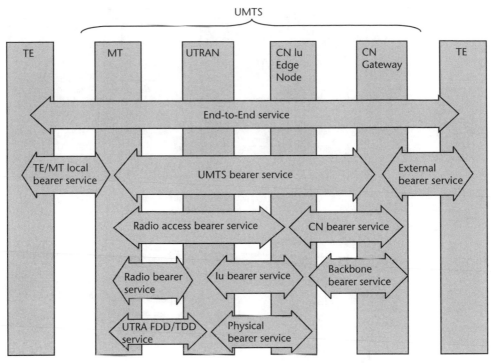

Figure 11.10 UMTS QoS architecture.

Table 11.2 UMTS Bearer Service Attributes

Attribute	Description
Traffic class	Conversational, streaming, interactive, and background classes, as with GPRS/GSM
Maximum bit rate	Maximum average bit rate, in the token bucket sense (i.e., the peak rate can exceed this, as long as the token rate is maximum bit rate)
Maximum SDU size	Size of token bucket (SDU is service data unit, the packet from the layer above)
SDU format information	Allowable formats (exact sizes) of SDUs
SDU error ratio	The desired maximum fraction of lost SDUs
Delivery of erroneous SDUs	About whether SDUs that are detected to have errors in them should be delivered or not
Guaranteed bit rate	Refers to a minimum guaranteed bit rate, especially useful for video streaming
Delivery order	Takes on the values "yes" if the bearer provides in-sequence SDU delivery, and "no" otherwise
Residual bit error ratio	Used for the error detection and correction algorithms on the wireless link
Transfer delay	The maximum delay tolerated (at least 95% of deliveries must fall within this value)
Traffic handling priority	Used only for the interactive class, this differentiates relative priority of traffic of that class
Allocation/retention policy	If network resources are scarce, traffic with a lower value of this attribute may not be admitted

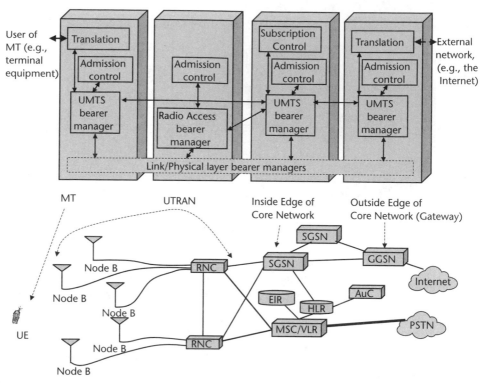

Figure 11.11   UMTS bearer service QoS management control architecture.

functions are found in the UMTS BS Manager, except for in the UTRAN, where they are in the RAB Manager. The service manager in each case interacts with admission and capacity control, the module that keeps track of all the available resources. Additionally, at the CN edge, there is subscription control that prevents users from successfully making requests for QoS services for which they are not subscribed. Clearly, in addition to these control functions, there need to be QoS mechanisms (as introduced in Chapter 7) in the user plane, so user data packets are handled appropriately. The arrangement of some of the mechanisms is shown in Figure 11.12. In the figure we see classifiers, traffic conditioners (Cond.), and mappers. The mappers work in conjunction with the classifiers by marking each packet appropriately for the lower-layer bearer, based on the higher-layer classification.

The QoS parameters also factor into the decisions made on whether to carry out certain procedures. For example, a QoS profile that specifies a low SDU error ratio may increase the chances that lossless serving RNS relocation is carried out, in the event of a serving radio network subsystem (RNS) relocation. What is serving RNS relocation? After soft handoff is completed between node Bs served by different RNCs (i.e., in different RNSs), the original RNC continues to be the anchor RNC even if the serving node B is under a different RNC. This leads to less efficient routing, and serving RNS relocation allows a switch of RNCs to the more

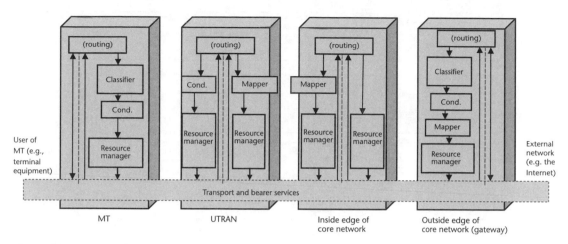

Figure 11.12 UMTS bearer service QoS management user plane.

appropriate one. Lossless RNS relocation is an option in RNS relocation that keeps track of packet sequence numbers so no packets are lost in the relocation process.

11.3.2 Security

UMTS security provides all the services of GSM security, plus some additional security services. In GSM, the network authenticates the MS, but the MS does not authenticate the network, so the system is vulnerable to fake BSs. Furthermore, confidentiality is only between the MS and BS in GSM. Communications within the GSM radio access network and core network are assumed to be secure enough and are not encrypted (ciphering only applies over the air). In UMTS, on the other hand, authentication has been extended so that the UE also authenticates the network. Encryption is also used within the wired portion of the UMTS network, unlike in GSM.

We can see some of the additional security features by looking at the overall UMTS security architecture as shown in Figure 11.13 [6]. Four security domains can be seen in the figure:

1. Network access security—partly found in GSM, but enhanced [including mutual authentication of UE and Node B, as mentioned, and also providing a data integrity service using a message authentication code (MAC) called f9], this is to secure underlying access to UMTS services, protecting especially against attacks over the wireless link.
2. Network domain security—new to UMTS, this secures communications on the wired portions of the network.
3. User domain security—available in GSM (SIM authentication), UMTS uses USIM authentication.

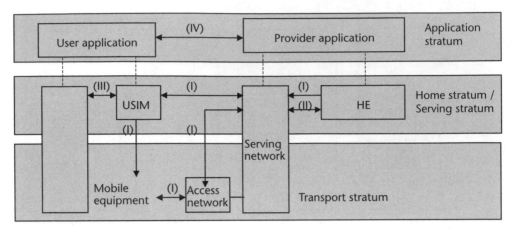

Figure 11.13 UMTS security architecture.

4. Application domain security—new to UMTS, this provides additional
   security for specific application domain communications.

Although the network access security is enhanced, it is similar in some ways to
GSM authentication, to make migration from GSM to UMTS systems easier. Thus, a
challenge-response scheme is still used. However, instead of just sending the random
number in the challenge, the network also sends an authentication token for network
authentication by the USIM. Another difference is that instead of obtaining authentication triplets from the HLR, the MSC or SGSN obtains authentication quintuplets.
Whereas the authentication vector (triplet) in GSM consisted of (1) random number,
(2) XRES, and (3) key to use for ciphering, in UMTS each vector also contains (4) the
authentication token and (5) the key to use for the MAC (for data integrity).

11.3.3 Different Releases

Since the mid-1990s, enhancements to GSM have been specified annually. These
enhancements have been specified in different *releases* identified by the year of
release, such as Release 1996 and Release 1997. Each release is a complete and consistent set of specifications for a cellular system. Later releases contain enhancements and new features not found in earlier releases. Specification documents
contain the appropriate release number in their title, so there can be no ambiguity
regarding what release a particular specification document is associated with.
Enhancements like GPRS, location services (using a location measurement system),
and intelligent network (see Chapter 10) features in GSM have first appeared in
these releases. Continuing this trend, UMTS has been introduced in a phased
approach, with different versions coming in almost annual releases, starting with
Release 1999.

However, what was to have been Release 2000 turned out to be too much to
handle within a year (there were too many work items), and so the work items were

split into two groups and designated to appear in two releases rather than one. At the same time, the naming of releases changed from being identified with a year to being identified with a sequence number. Thus, the new releases became known as Release 4 and Release 5. They were completed in March 2001 and June 2002, respectively. Release 6 is nearing completion as of this writing.

The major features of Release 1999 include:

- New radio access network, UTRAN, featuring WCDMA;
- Enhanced data rates for GSM evolution (EDGE), an alternative to WCDMA that increases the capacity of GSM time slots;
- Enhanced messaging services, including multimedia messaging service (MMS);
- Advancements in many aspects of the core network, including the intelligent network features, security architecture, and GPRS;
- Introduction of OSA.

The major features of Release 4 include:

- More complete support of location services (LCS);
- Changes in UTRAN transport to better support IP;
- Introduction of robust header compression (Chapter 4);
- Improvements in messaging services;
- Bearer-independent circuit-switched domain architecture (allowing the possibility of a common IP transport in the CN, as will be further discussed in Chapter 12);
- Evolution of OSA.

The major features of Release 5 include:

- Enhancements to LCS;
- Further changes in the UTRAN to allow IP transport as an alternative to ATM transport (in Release 1999 and Release 4, only ATM transport is allowed);
- Network domain security is introduced in the UMTS CN;
- Introduction of high-speed downlink packet access (HSDPA), a new feature that provides instantaneous data rates of up to 10 Mbps in the downlink, to support services like video on demand;
- RNCs are allowed to connect to more than one MSC and more than one SGSN, the primary benefit being better load balancing between MSCs or SGSNs;
- Introduction of the IP Multimedia Subsystem (IMS), which will be discussed in Chapter 12;
- Evolution of OSA;

- Introduction of wideband adaptive multirate (wideband AMR) codec, which will provide higher speech quality than traditional GSM codecs provide;
- Enhancements in end-to-end QoS support;
- Improvements in messaging services.

The major features of Release 6 include:

- Addition of MIMO techniques at the physical layer for higher data rates;
- Enhancements to LCS, including enhanced support for anonymity and user privacy;
- IMS (more on IMS in Chapter 12) phase 2, including policy-based QoS control, and IMS support for conferencing, messaging, local services, and emergency calls;
- Support of presence capabilities;
- Multimedia broadcast and multicast services (MBMS);
- Interworking with WLAN;
- Continuing enhancements on OSA;
- Improvements in MMS.

More detailed information can be found on the 3GPP Web site [7].

11.3.4 Comparison with cdma2000 Development

UMTS and cdma2000 are similar in many ways. In order to achieve higher data rates, both 3G systems use wider radio bandwidths for their channels. It is true that in the meantime certain recent enhancements to 2G systems have been made so that they obtain higher performance, such as higher data rates, within their existing bandwidth. However, the enhancements are limited by constraints arising from the physics of how much could be squeezed into limited radio bandwidth. Examples of these enhancements to 2G systems are EDGE for GSM and the 1X version of cdma2000 for IS-95, where 1X refers to the fact that it uses the same radio bandwidth as IS-95, whereas 3X and 5X use three and five times the radio bandwidth of IS-95, respectively. Also, both systems use similar fundamental techniques for channelization of the wireless medium, based on direct-sequence spread spectrum.

There is a consensus that wireless cellular systems should move more towards IP-centric core networks. As designers for both systems looked forward to the need to eventually provide IP multimedia services using IP-centric control and management protocols, they ended up working on solutions that are very similar. Both use SIP for much of the related signaling. Both introduced new network elements where various functions, such as keeping track of call state, were clustered, and the grouping of these functions in the two systems is quite similar (even if the network elements do not have the same names). In Chapter 12, we will explore the IMS of UMTS in more detail.

However, there are important differences between UMTS and cdma2000. Many of these differences arise from the fact that UMTS is designed with backward compatibility with GSM in mind, whereas cdma2000 is designed with backward compatibility with IS-95 in mind. Backward compatibility does not mean that all equipment (e.g., BSs, handsets) for the new system always works with all equipment (e.g., BSs, handsets) for the old system. However, it does imply that the old and new systems have some elements in common. For example, the new system may be designed such that its clock rate is a multiple of that of the old system. Thus, a dual-mode phone supporting both systems could use a common clock (as well as other common elements) and hence be cost-effective. The backward compatibility of cdma2000 with IS-95 is stronger because both are based on CDMA technology.

On the network side, we see more examples of backward compatibility, as network protocols from the corresponding 2G systems have been evolved, adapted and enhanced. We notice that in certain cases where some functionality existed in 2G systems, and it was desired to have similar functionality in the evolved 3G systems, the solutions chosen by UMTS and cdma2000 differ the most. On the other hand, where some functionality needed in 3G systems is completely new and does not have a counterpart in the 2G systems, the solutions chosen by UMTS and cdma2000 are strikingly similar. For example, in the case of packet-data support, some functionality existed in the 2G systems, so UMTS retains GPRS with some modifications, whereas cdma2000 adopts a more "pure" IP-based solution. Many aspects of mobility management are handled by GPRS and by Mobile IP, respectively. In the case of IP multimedia service control and management, on the other hand, the functionality was lacking in the 2G systems. Thus, both UMTS and cdma2000 needed new subsystems of network elements to handle multimedia service control and management. This resulted in similar solutions.

In the race between UMTS and cdma2000 towards commercial deployment, the cdma2000 camp has seized the early lead. One reason for this is that there are more similarities between cdma2000 and IS-95 than between GSM and UMTS, so one would expect fewer hurdles to overcome. Another major reason is that IS-95 systems have evolved to two high data rate technologies that still use the original IS-95 spectral bandwidth, the 1X Evolution-Data Optimized (1XEV-DO) and 1X EVolution-Data and Voice (1XEV-DV) technologies. Neither 1XEV-DO [also known as high data rate (HDR)] nor 1XEV-DV is an earlier version of the other, but they could be thought of as systems resulting from two different evolutionary paths from IS-95, with some similarities but optimized for different applications (data only, e.g., for asymmetric Internet browsing and video streaming, in the case of 1XEV-DO, and both data and voice, in the case of 1XEV-DV). Some proclaim that these qualify as 3G technologies, and that therefore 3G has touched down first with the cdma2000 camp. Meanwhile, UMTS deployment has been slow, but is also gaining momentum. The first UMTS system to be commercially deployed was the Freedom of Mobile Access (FOMA) system by NTT DoCoMo in Japan in 2001. After some sluggishness in 2002, the rate of deployments picked up in 2003 and 2004.

11.4   Summary

While the Internet has been adapting to new requirements, including support for wireless links and mobility, the convergence of wireless and the Internet has a second major face. Traditionally circuit-switched, telephony-based cellular systems are evolving towards more accommodation of packet-switched IP-based communications; many would say that the evolution will end in the so-called all-IP wireless network. Even if the end of the evolution is not a "pure" all-IP network, the changes in cellular systems are still of great interest, and enlightening and pertinent to the subject of this book. Thus, this chapter surveys the evolution of 2G systems to 3G, focusing on the evolution from GSM to UMTS. We analyze how GPRS was added to GSM as an intermediate step, to better support data traffic. We discuss the drivers for moving towards 3G systems, and then introduce UMTS. Since cdma2000 is the other major 3G system, we also compare UMTS with cdma2000.

References

[1]   http://www.gsmworld.com.

[2]   http://www.umts-forum.com, "GSM: The Business of a Billion People," UMTS Forum Press Release, February 23, 2004.

[3]   3GPP TS 23.060 v.4.9.0 (2003–12), "General Packet Radio Service (GPRS); Service Description; Stage 2 (Release 4)," December 2003.

[4]   3GPP TS 23.107 V3.9.0 (2002-09), "Quality of Service (QoS) Concept and Architecture (Release 1999)."

[5]   Holma, H., and A. Toskala, *WCDMA for UMTS*, 2nd ed., New York: John Wiley & Sons, 2002.

[6]   3GPP TS 33.102 V3.13.0 (2002-12), "3G Security; Security Architecture (Release 1999)," December 2002.

[7]   http://www.3gpp.org.

CHAPTER 12

The IP Multimedia Subsystem (IMS)

In Chapter 10, we discuss general trends in the convergence of wireless and IP. We have seen how, in moving towards all-IP networks, network providers would like to provide more than just IP connectivity. In versions of UMTS from Release 5 onwards, provision of basic multimedia over IP services will be handled by a new network subsystem, the IMS [sometimes also known as IP multimedia core network subsystem (IM CN SS)]. These basic multimedia over IP services include the establishment, management, and termination of voice, video, and other sessions carried over IP. Additionally, IMS takes a layered and flexible approach to service provision, by providing hooks for application servers to offer further services built upon the basic service capabilities.

This chapter provides a good example of how wireless and IP come together in a real telecommunications system, to provide basic and advanced telecommunications services. It ties together elements from all the other chapters in the book so far, and illustrates some of the system design choices involved in wireless IP telecommunication system design.

Since IMS is a new, evolving subsystem, we base this chapter on the UMTS Release 5 version of IMS.

12.1 Motivations and Requirements

In moving towards an all-IP network, the evolving releases of UMTS continue to provide voice services. One reason for this is that despite the growth in volume and diversity of data traffic, voice continues to be important for the foreseeable future because it supports such a basic and fundamental form of human communication. Furthermore, the growth and enhancement of a familiar product is less risky than a complete redefinition of the product—subscribers of mobile systems are more likely to remain loyal customers if they perceive the new systems to be enhancements of the ones they are familiar with than if new and different services are available. It is more desirable to build on the existing customer base than to find a new base from scratch.

Thus, both voice and data should be supported. To continue to provide voice services while at the same time adding packet-switching support for data, and IP connectivity in particular, two approaches could be adopted. First, two separate networks could be developed for circuit-switched voice and packet-switched data. Second, an integrated network could handle both types of traffic. The first approach is taken with GPRS, where the GSNs form a new core network-subsystem different from that formed by the MSCs, and parallel to the circuit-switched infrastructure. However, it is an interim solution that is not so efficient or cost-effective, as two parallel networks need to be maintained. In the second approach, the first question to ask is whether the integrated network should be circuit switched or packet switched. Both circuit-switched and packet-switched networks can handle both voice and data (voice over packet-switched networks is explored in Chapter 4). Several factors point in favor of packet switching, however. As explained in Chapter 4, the bursty nature of data traffic makes it more efficient to use packet switching for data. Thus, from a resource utilization perspective, circuit switching it is not acceptable (although virtual circuits may be okay if efficient resource utilization technologies are available, as in the case of ATM). Thus, the increasing dominance of data traffic makes for a volume argument that the network should use the switching technology best suited for the bulk of the traffic, which is packet switching. Furthermore, the rise of the Internet makes IP the packet-switching technology of choice, whatever the underlying transport mechanism, such as IP over ATM or IP over frame relay.

To achieve the all-IP network as an evolution of GSM, three stages of development can be identified, which are logical steps towards the all-IP network. These are:

1. Provide IP connectivity for data traffic, with a packet-switching core network and efficient sharing of GSM time slots for packets over the air interface.
2. Switch over internal voice traffic transport to the internal IP network, keeping existing legacy telephony-style control protocols, such as GSM MAP (as we have seen in Chapter 3, Section 3.2.1).
3. Switch over control (protocols and signaling) to IP-style protocols.

Table 12.1 summarizes the three steps just described and shows when they were accomplished in the evolution of GSM to UMTS.

Why would there be a need to switch over to IP-style protocols? Because these are IP-style protocols, they integrate easier and more completely with other IP-style protocols for such functions as Web browsing and e-mail. Many new services and applications could be developed as a result. Furthermore, the IP-style protocols do not need to maintain features of the old protocols that are not relevant to an IP

Table 12.1   Major Steps in Moving Towards the All-IP Network

Step	Accomplished
IP connectivity for data traffic	With introduction of GPRS
Switch over internal voice traffic transport to internal IP network	With UMTS Release 4
Switch over control (protocols and signaling) to IP-style protocols	Ongoing work in UMTS (started with Release 5)

environment. Refer to Chapter 5 for more discussion on these points. Step 3, switching over to IP-style protocols, requires careful thinking because of the complexity of the signaling required, which must handle all the QoS, security, and mobility issues that the old circuit-switched protocols used to handle. While some of the functions are similar to, and parallel, those found in the circuit-switched case, there are some differences due to variations between packet-switched and circuit-switched traffic. UMTS has introduced a whole new network subsystem, the IMS, with new network elements that use IP-style protocols to provide call control services for multimedia over IP sessions. We will discuss the architecture of the IMS in Section 12.2.

The first question that one naturally asks is whether it is necessary for the UMTS network itself to provide these services. Why not just provide IP connectivity and let the users freely use the network? Some may argue that the network operators should not be providing high-level services, but just network-level services like IP connectivity, for the simple reason that it is notoriously hard to predict what services will be successful and popular among customers (as seen in Chapter 10). The IMS handles this issue in a two-pronged manner. First, the IMS provides basic services like session establishment (of multimedia over IP sessions) that can be used both as high-level services by the subscriber (e.g., when a subscriber makes direct use of session establishment to make a video call), as well as mid-level services for composing other high-level services (e.g., when an application makes use of session establishment capabilities as part of providing high-level services to subscribers). Second, it provides hooks and means for application development using the services of IMS.

The second question that one naturally asks is what control signaling protocol to use for the multimedia over IP services. It would need to replace the legacy telephony-style control signaling protocols, be flexible, and be easy to integrate with other IP protocols. As we have seen in Chapter 5, SIP meets these requirements, and so it is the prime candidate to be the control signaling protocol for the multimedia over IP services that UMTS will provide.

12.1.1   Using SIP

Of course, one could suggest other alternatives to using SIP, such as developing another signaling protocol from scratch, or modifying something else, such as the H.323 protocol suite. Developing another signaling protocol from scratch would set back development efforts by years, as the new protocol is defined, refined, and tested. Furthermore, it would be unnecessary duplication of effort, since SIP, H.323, and other protocols already provide some of the desired capabilities. It would make more sense to modify one of these existing protocols and adapt it to the particular needs of UMTS for providing multimedia over IP services. In fact, this would be similar to the approach used in designing GSM MAP—it is an extension to the well-established SS7 protocols and is carried on SS7 signaling networks between PLMNs—rather than a completely new protocol. In choosing an appropriate protocol for control signaling, SIP may be the most suitable candidate, since it is lightweight (not encumbered with lots of unnecessary features) and flexible.

Moreover, it is text-based, with semantics that bear a family resemblance to other IP-style protocols like http (as we had seen in Chapter 5), supporting flexible and powerful integration of IP-based services.

However, there are some deficiencies of SIP that need to be addressed before it can be used in a UMTS network environment. More precisely, there were these deficiencies when 3GPP began working on UMTS Release 5. During the course of the standardizations process, 3GPP delegates suggested a variety of enhancements and additions to SIP and to the ways it could be used. The major changes have been fed back into the IETF so that SIP itself has been changing to support some of the features required by its use in UMTS.[1] Examples will be discussed in the exposition in the text as we proceed in this chapter. These examples include:

- At first, SIP did not provide a way for a proxy to insert itself into the signaling path to or from a SIP user for the duration of a registration; there was only a way, using the record-route option, for a proxy to insert itself for the duration of a session (not the duration of a registration). Thus 3GPP proposed the Path header to be added to the SIP header, and later this was incorporated into the official SIP specifications [1].

- Certain sessions require some conditions to be met in order to be successfully established (see the discussion on QoS requirements in Section 12.1.2, for instance). For example, end-to-end resource reservation for QoS might need to be ensured before a session can be successfully initiated. Individuals from 3GPP involved in IMS development were among the developers of SIP extensions to provide for this need. A new SIP method (the UPDATE method) was developed, and a mechanism was also developed making use of that method to ensure successful resource reservation before ringing of the phone [2, 3].

- In a SIP signaling flow, there are two kinds of responses, namely final responses and provisional responses. The official response to a SIP request is a corresponding final response, but before the final response is sent, one or more provisional responses may be sent (as explained in Chapter 5). Since these were originally intended to be informational responses (e.g., before a final response to an INVITE is sent, provisional responses could be sent indicating "ringing" is occurring), reliable transmission is not required. Therefore, no acknowledgement is required for provisional responses. However, in certain cases where SIP is used in operator networks like UMTS networks, especially to support PSTN interworking, some provisional responses must be reliably transmitted, because certain call-state-machine transitions depend upon their receipt. Thus, SIP was extended to allow acknowledgments for provisional

---

1. We may recall the principle from Chapter 1 that things are often used in ways for which they were not originally designed, and thus the design should be flexible to allow for such adaptations. So far, SIP appears to be holding up pretty well in the face of various expansions in usage and new requirements in new usage scenarios.

responses. This new type of acknowledgment is called a provisional response acknowledgment (PRACK) [4].

Moreover, in a commercial network like UMTS, there are other considerations that lead to SIP being used in certain ways in UMTS that may be different from its use in a hobbyist environment (e.g., end-to-end signaling directly between friends), but that do not need changes to SIP itself. In other words, SIP is flexible enough to be used in these ways, even if these are new architectures, in terms of arrangement of servers and the roles of SIP servers. For example, we look at the call state control functions (CSCF) in IMS. CSCF is the UMTS terminology for a SIP server with certain roles to play in IMS, and there are three kinds of CSCFs used in IMS. The serving CSCF (S-CSCF) plays the role of the home SIP proxy that knows where the MS is, since it also doubles as the home SIP registrar for the MS. There can be multiple S-CSCFs in a network, for load balancing and other purposes. Since a network operator may wish to hide its network configuration from outsiders, it may not want SIP messages entering its network to be addressed directly to the S-CSCFs. Thus, the interrogating CSCF (I-CSCF) is used as the entry point to the network, so S-CSCFs and the rest of the network can be hidden. While network hiding is an optional feature, the I-CSCF also has a nonoptional role, to discover the home S-CSCF of a mobile terminal.

Furthermore, since for many years to come, there will still be phones on traditional circuit-switched phone networks (the PSTN), it would be necessary for calls to be made between UMTS subscribers and such PSTN phones. Since different protocols are used on the two sides, the choices are: (1) to force the use of one of the two protocols for these intersystem calls or (2) to interwork the two protocols at special gateways that translate between one protocol and the other (both ways) so a different protocol can be used on each side of the gateway. Choice (1) is impractical: Clearly the PSTN side could not be expected to be revamped to understand SIP just to support such intersystem calls, so it would be the UMTS network that would be forced to use telephony-style protocols, meaning it would need to maintain circuit-switched elements and thus lose one of the main benefits of an integrated packet-switching network. Thus, choice (2) is the right way to go.

In summary, then, as a result of the decision to use SIP as the session-control protocol in UMTS for multimedia over IP services, SIP itself was found to lack certain features needed for its use in UMTS, and some of these changes found their way back into the SIP specifications and related specifications in the IETF. In addition, SIP needs to be used in a particular, controlled way, and interworking with the PSTN needs to exist. In Sections 12.2 and 12.3, we will see how these requirements are met in the IMS.

12.1.2 Other Requirements

In addition to the requirements directly related to the use of SIP that we just discussed, other requirements are placed on the IMS by 3GPP.

Perhaps the most interesting of these requirements is to use IPv6 in the IMS. It is a forward-looking requirement, since IPv4 is still the dominant version of IP in the world today. Since there is less experience in IPv6 networks among vendors and engineers in general, the decision to mandate IPv6 in IMS may cost a few extra hiccups in implementation of IMS. However, it may be the right choice in the long run, especially once IPv6 takes off. UMTS network operators will not have to worry about when and how to upgrade their IMS domains to IPv6.

Since IMS sessions make use of GPRS transport, the GPRS security features are used for GPRS network access control, encryption, and other functions. On top of that, access control mechanisms for the IMS domain subnetwork, such as implementation by gateways, are required. Since the SGSN and S-CSCF are different network elements, the GPRS mechanisms for secure data transfer between MS and SGSN are insufficient to provide for secure communications between MS and S-CSCF, and additional schemes are needed for this. Also between the network elements in the IMS domain, secure communications are needed. Another aspect of security is that the hooks for allowing third-party access to IMS (to develop applications that build on the basic capabilities) should provide only for controlled, authenticated access to the network capabilities. This is not a problem with OSA, since OSA is designed to support such controlled access.

QoS for IMS sessions is built upon the UMTS-layered QoS framework. In addition, to allow QoS signaling and session-control signaling to each evolve with minimal impact to the other, the QoS signaling and session-control signaling must be independent. However, end-to-end QoS negotiation and resource allocation must occur in conjunction with session establishment (thus, certain sessions could fail to be set up because of failure to meet their QoS demands). In particular, the calling party may require certain resources to be allocated successfully before the called party is alerted. Moreover, since IMS signaling traffic is more important than session traffic, the network should be able to give preferential treatment to signaling traffic.

Finally, the IMS needs a variety of multimedia-stream processing and handling capabilities, such as the ability to mix two or more streams for multiparty sessions and perform transcoding of streams and analysis of streams. Capabilities for sourcing multimedia streams are also needed, for example, for playing multimedia announcements.

12.2 IMS Architecture

IMS consists of the network elements that provide the SIP-based session control for the new IP multimedia services that UMTS will support. Before getting into the specifics of how these network elements are used, we discuss a few overall features of the IMS architecture.

In Chapter 11, we have seen how the GPRS network was added to GSM, and then carried over to UMTS as the network evolved. Thus, a packet-switched (PS) domain was added to the old circuit-switched (CS) domain, where the PS domain

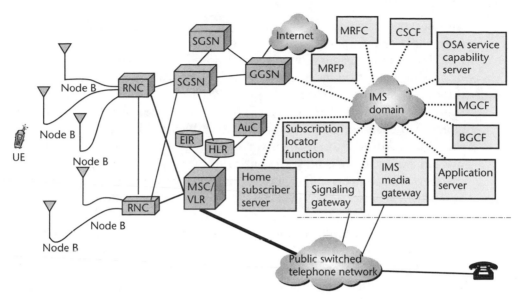

Figure 12.1 IMS.

includes network elements like the SGSNs and GGSNs, whereas the CS domain includes the MSCs and others. IMS is a third domain added to UMTS. The network architecture is shown in Figure 12.1. We will discuss the new network elements in Section 12.2.2.

The IMS security architecture is a layered architecture similar to the more general UMTS security architecture seen in Chapter 11, but with the addition of security associations between IMS network elements. Thus, it goes over and beyond GPRS security, as required, by providing for security associations between the MS and P-CSCF, and MS and S-CSCF.

12.2.1 Relationship with GPRS

IMS is both strongly dependent on GPRS and independent of GPRS. There can be UMTS networks with only a CS domain, or UMTS networks with only a PS domain, but not UMTS networks with only IMS. IMS only makes sense in a network with a PS domain, and specifically with GPRS. IMS relies on GPRS to deliver SIP signaling messages to and from MSs. GPRS gives the appearance of a virtual link between an MS and a corresponding entry point to IMS, in other words, a P-CSCF. In addition to the use of GPRS to transport IMS signaling messages between an MS and the IMS network subsystem, GPRS is also used to transport the IMS multimedia session traffic. IMS multimedia session traffic is the voice, video, or other data packets that are part of the sessions established by IMS signaling. Note that there is no reason for this traffic to flow through the IMS network subsystem, nor is it desirable to do so—this is the principle of separation of signaling and data paths, which we first saw in Chapter 5. For example, for a session between an MS and a host on the

Internet, once the session is set up, the session traffic can go directly from the MS over GPRS to the appropriate GGSN to the Internet.

Having just seen the ways in which IMS is strongly dependent on GPRS, we now discuss how it is also independent of GPRS. Instead of extending GPRS to include the new IMS network elements, the 3GPP delegates chose to create a new UMTS subnetwork for IMS. This was the right choice, in the spirit of modular design principles. Although both are IP networks, the GPRS and IMS networks may each have their own address space. Also, as we have already mentioned, IPv6 is mandatory for IMS. However, there is no such constraint for GPRS, and thus it is quite likely that an operator's network might have an IPv4-based GPRS network next to an IPv6-based IMS network. Also, IMS does not solely rely on GPRS security mechanisms but adds mechanisms for security associations between the MS and S-CSCF.

Note that both IMS signaling traffic and the corresponding multimedia session traffic are transported by GPRS as user data (using the GPRS user plane protocol stack). However, in order to satisfy different QoS requirements for signaling and media traffic, different PDP contexts are used for signaling and media traffic. In fact, before any IP multimedia services can be provided, a PDP context must be activated for carrying IMS signaling. Later, when multimedia sessions are established, one or more new PDP contexts (possibly secondary PDP contexts) would be activated for these streams, depending on their QoS requirements.

12.2.2 Network Elements

The main network elements in IMS are the CSCFs, the media gateway control function (MGCF), the media gateway, the multimedia resource function controller, the multimedia resource function processor, subscription locator function, breakout gateway control function, application server, signaling gateway function (SGW), and security gateway. Additionally, there is a network element called the home subscriber server (HSS) that can be thought of as an augmented home location register (HLR) for circuit-switched services, which is not just an IMS network element, but also is accessed by circuit-switched domain elements like the MSC. We have already introduced the three types of CSCFs in Section 12.1.1. We will see in Section 12.3 how they are involved in signaling to meet requirements for registration and call setup in various scenarios. Other requirements, as discussed in Sections 12.1.1 and 12.1.2, include the following: capabilities to interwork with the PSTN, media mixing and processing capabilities, hooks for higher-level service creation, and security policy enforcement. We now discuss how the IMS network elements work together to meet these requirements.

Interworking with the PSTN uses the MGCF, media gateway, breakout gateway control function, and SGW. For calls between an IMS subscriber and a PSTN subscriber, the voice packets are carried over IP on one side (media streams) and over voice circuits on the other side, so there must be a point where a conversion takes place. This point is the media gateway. A variety of codecs may need to be supported in a media gateway, including other processing capabilities like echo canceling.

Rather than having the control functions for the media gateway co-located with the media gateway, they are located separately in the MGCF (Media Gateway Control Function). You may recall that we have described this kind of separation before, in Chapter 4. An additional function that is not discussed in Chapter 4 is the breakout gateway control function. It arises because there could be multiple MGCFs that could be the point where the call "breaks out" from the UMTS network into the PSTN. For example, if an IMS subscriber is making an international phone call, it may be better (perhaps more economical) for the IP media streams to go all the way to the destination country (or at least, part of the way) before breaking out to the PSTN, than to break out in the country from where the IMS subscriber is calling. However, the feasibility of doing so may depend on other factors such as interoperator agreements. Thus, the breakout gateway control function is assigned the task of selecting an appropriate MGCF where breakouts should occur. It can do this on a call-by-call basis.

The multimedia resource function controller and multimedia resource function processor work together to provide a variety of multimedia-stream processing and handling capabilities, including mixing, sourcing, and transcoding, as mentioned in Section 12.1.2. Thus, they provide support for multiparty sessions, playing of announcements, and other such useful features. The division of labor between the controller and processor is similar to that between media gateway controllers and media gateways, following the modular model of separating control and processing. This allows a controller to control multiple processors, and for some independent upgrades to either element.

12.3 IMS Procedures

12.3.1 Registration

In order to be able to provide service to a subscriber, the IMS needs to know how to reach the user. An S-CSCF assigned to the subscriber in his or her home network, and a suitable P-CSCF should be assigned in the network currently serving the subscriber. Note that even though the MS may be IMSI-attached and GPRS-attached, the MS location would be known only in the CS and PS domains. The IMS does not know where the MS is, and the MS is not yet set up to use IMS services. So why does not, for example, the GGSN inform the IMS domain when an MS becomes GPRS-attached? First, the GGSN does not have all the relevant IMS-related information on the MS, and second, the MS may wish to use only GPRS services and not IMS services. It would seem best if the MS itself handles the process of setting itself up for IMS services, and indeed, this is the approach taken. Registration is the IMS procedure to meet these requirements. Before registration can be carried out, GPRS attach, PDP context activation, and P-CSCF discovery need to occur, since it builds on these functions, as shown in Figure 12.2.

IMS registration is basically a SIP registration by the MS (and the basic, simplified idea is shown in Figure 12.3), although there are some interesting aspects to the

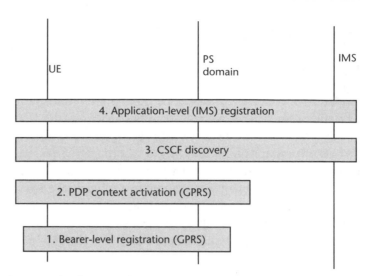

Figure 12.2 Sequence leading to IMS registration.

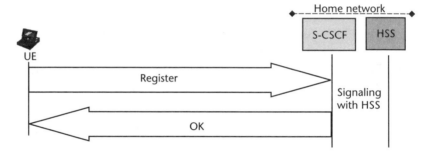

Figure 12.3 Registration, simplified.

way it is done. The P-CSCF ensures that it is in the signaling path for SIP signaling to and from the MS for the duration of the registration, so it can handle matters like resource allocation during session initiation. The P-CSCF puts itself into the signaling path by using the Path header in the SIP REGISTER message (recall that the Path header was not originally in SIP, but was later added to meet requirements such as this usage in IMS). Another feature is that the SIP registrar is dynamically chosen by the I-CSCF in conjunction with the HSS. In particular, the S-CSCF acts as a SIP registrar, but there is no fixed assignment of S-CSCF to IMS subscribers. This allows features like load balancing to be provided.

The registration process is as follows: the MS sends a SIP REGISTER to the P-CSCF (it would have discovered the P-CSCF through one of several ways, including having obtained the information during PDP context activation). The P-CSCF discovers the I-CSCF (the entry point to the subscriber's home network) by examining the information in the REGISTER message, and forwards the message there.

The I-CSCF queries the HSS for information like S-CSCF capabilities and current loading of candidate S-CSCFs, to decide which S-CSCF to dynamically assign to the MS. The REGISTER message is then forwarded to this S-CSCF, which acts as a SIP registrar, storing the relevant information. The S-CSCF also communicates with the HSS to confirm that it is serving the MS and to obtain pertinent subscriber profile information. This communication with the HSS does not use SIP but some other protocol. The S-CSCF then responds to the MS with a SIP 200 OK message.

Before registration, the P-CSCF in the subscriber's serving network has no knowledge of the subscriber (not the MS address, or its user IDs, or how to reach its home network), no S-CSCF is assigned to serve the subscriber, and the subscriber's home network IMS does not know how to contact the subscriber. All this is taken care of during registration. Thus, after registration, the P-CSCF knows the MS address, its user IDs, and how to reach its home network. An S-CSCF is assigned to the user that knows the user IDs and IP address, some subscriber profile information, and how to reach the P-CSCF, and the HSS knows which S-CSCF is assigned to the user.

12.3.2 Basic Wireless-to-Wireless Call

We now describe the IMS session initiation procedure. You may like to bear in mind the traditional GSM call setup signaling (as described in Chapter 3), as well as the traditional SIP session initiation signaling (as described in Chapter 5). You will find similarities with both. IMS session initiation uses SIP, but in a particular way, taking into account specifics of the wireless environment (where one or both parties could be roaming instead of at home), interworking with the PSTN, operator issues, and interoperator issues. We have already discussed some of these requirements in Section 12.1.1 and we now see an illustration of how the requirements are met in the session-initiation procedure.

We begin with a basic wireless-to-wireless call where both parties are IMS subscribers and both are in their respective home networks. On both sides, the signaling from the MS goes through their respective P-CSCFs and arrives at their respective S-CSCFs. The portion of the signaling between the two S-CSCFs is shown in Figure 12.4. There is an optional I-CSCF in the signaling path in the figure. This is for the case that the originating and terminating operators are different *and* the originating network wishes to hide its topology from the other network [thus, the I-CSCF serves as a topology hiding interworking gateway (THIG)]. In the cases that the same operator is serving both MSs, or that they are different and the originating network does not hide its topology from the other network, that I-CSCF is not used.

We note that the signaling flow is similar to a regular SIP signaling flow with some peculiarities. The presence of the service-control boxes in the flow acknowledge the fact that either operator may invoke certain service logic (e.g., to play an announcement) at certain points in the signaling flow. The location query and response between I-CSCF and HSS may remind you of the use of a SIP redirect

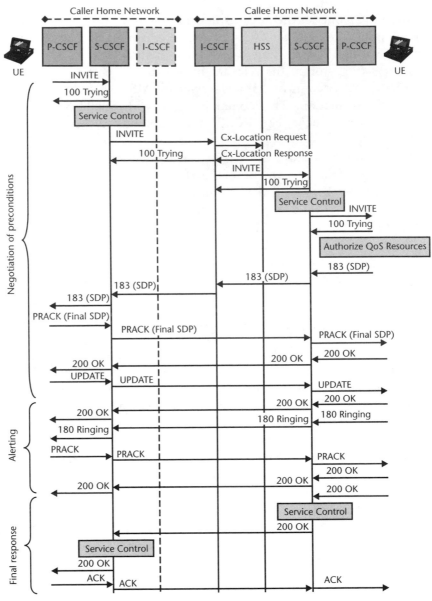

Figure 12.4 Signaling flow for basic session initiation.

server, and in fact the HSS does act as a redirect server to help the I-CSCF locate the S-CSCF of the called party. Then, as you did not see in Chapter 5, there are a number of exchanges (offer response, response confirmation, confirmation acknowledgment, reservation confirmation, and reservation confirmation). This is the negotiation of preconditions (QoS) stage that happens before ringing, as we mentioned earlier. The SIP UPDATE message could be used for reservation confirmation. Finally, ringing occurs, followed by pickup on the called-party side.

12.3.3 Changes for Roaming Caller or Roaming Called Party

Now, suppose that either the caller or the called party, or both, are roaming. What difference does this make to the session initiation signaling? First we consider the case that only the caller is roaming but the called party is in its home network. Then we consider the case that only the called party is roaming but the caller is in its home network. Finally, we consider the case that both sides are roaming.

If only the caller is roaming, the signaling flow is very similar to the home-to-home case we saw in Section 12.3.2. However, the P-CSCF of the caller would be in a roaming network, not in the home network. So, there would need to be a mechanism to map user identities to the well-known address of the appropriate I-CSCF of the caller in the associated home networks. From the I-CSCF onwards, the signaling is the same as in the case that the caller is in its home network. However, the S-CSCF may use different criteria for allowing the session to be initiated than in the case when the subscriber is at home. For example, while the S-CSCF may allow subscribers in their home network to originate sessions most of the time (with only a few exceptions, such as when the subscriber is using prepaid service and has a very low balance), other criteria may apply in roaming situations. Note that the roaming network may wish to hide its network topology from other operators, and so it may optionally pass the signaling from the P-CSCF through an outgoing I-CSCF on the way to the I-CSCF of the caller's home network.

If, instead, the called party is roaming, the difference comes in the portion of signaling between the S-CSCF of the called party and the P-CSCF of the called party. The S-CSCF would need to locate the P-CSCF in the appropriate roaming network, using the latest registration information from the called MS. As in the case of the roaming subscriber originating a session, in this case certain factors may factor into the decision of the S-CSCF to allow the signaling to proceed to the P-CSCF and the roaming MS; these factors may be different from the case in which the MS is not roaming. Assuming the S-CSCF decides to let the session origination proceed, the SIP INVITE message will go to the appropriate P-CSCF in the roamed-to network. Again, if the home network wishes to hide its topology, the signaling will go out through an I-CSCF in the home network. Similarly, if the roaming network wishes to hide its topology, it will add its own I-CSCF in the signaling path.

In the event that both the calling and called parties are roaming, it is straightforward simply to combine the two sets of modifications that we have just discussed, since each just affects one side of the signaling. Thus, there could be up to four networks involved: the home and roaming networks of the caller, and the home and roaming networks of the called party. If all these networks wish to hide their topology, there could be six I-CSCFs involved, including two from each home network.

12.3.4 Changes for Wired Caller or Wired Called Party

If the caller is using a phone on the PSTN, and the called party is a UMTS subscriber, signaling will be routed to an MGCF in the home network of the called party. Once the signaling reaches the MGCF, it is translated to the relevant SIP signaling, from the ISUP signaling used in the PSTN, as shown in Figure 12.5.

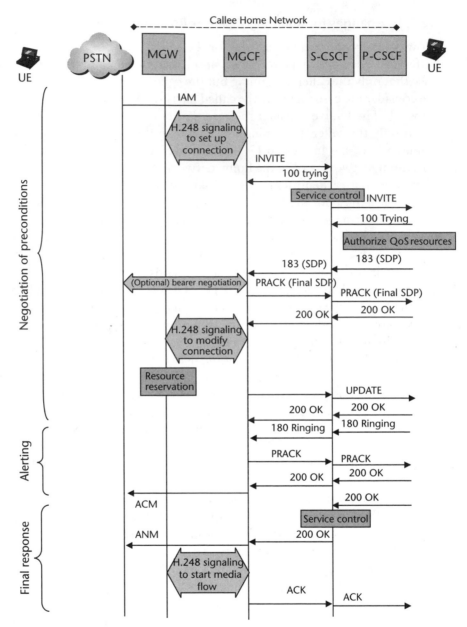

Figure 12.5   PSTN origination.

If it is the called party that is in the PSTN (Figure 12.6), then the BGCF in the home network of the calling party is responsible for locating the breakout point to the PSTN. Thus, it can decide to do the breakout in the same network, whereupon it would forward the signaling to the MGCF in the same network, or it can decide to do the breakout in another network, whereupon it would forward the signaling to a BGCF in the other network, to forward to an MGCF in that other network.

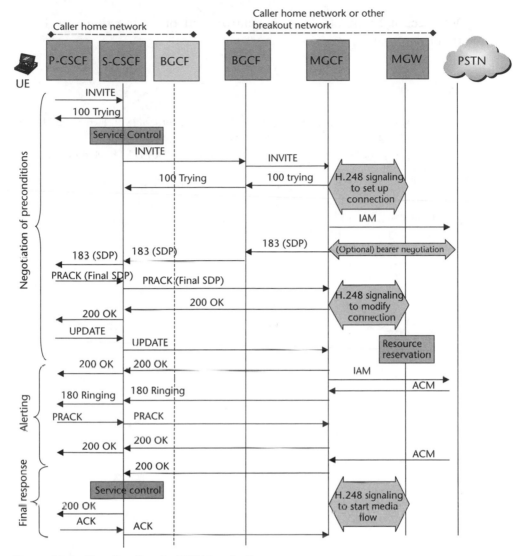

Figure 12.6 Signaling flow for PSTN termination.

12.4 Summary

We see in Chapter 11 that UMTS is not just one system, but comes in different releases, the later releases building upon the earlier releases to incorporate more enhancements and features. While IP transport was added to UMTS in earlier releases, a major addition in UMTS release 5 was the IMS. IMS allows IP-style session and call-control signaling (with SIP) to be used in UMTS, and we explore its basic features in this chapter. UMTS with IMS forms an interesting case study of a real wireless Internet system, and thus in Chapter 11 and this chapter, we see how

QoS, security, mobility, session control, and other functions are handled in a complete system.

References

[1]   Willis, D., and B. Hoeneisen, "Session Initiation Protocol (SIP) Extension Header Field for Registering Non-Adjacent Contacts," RFC 3327, December 2002.

[2]   Rosenberg, J., "The Session Initiation Protocol (SIP) UPDATE Method," RFC 3311, September 2002.

[3]   Camarillo, G., W. Marshall, and J. Rosenberg, "Integration of Resource Management and Session Initiation Protocol (SIP)," RFC 3312, October 2002.

[4]   Rosenberg, J., and H. Schulzrinne, "Reliability of Provisional Responses in the Session Initiation Protocol (SIP)," RFC 3262, June 2002.

CHAPTER 13

Future Possibilities

13.1 What Is 4G?

Technology keeps progressing. Perhaps in 10 to 20 years from now, today's 3G mobile systems will be perceived as using old technology, unable to provide the services needed by new and upcoming applications. Such perceptions drove the move from 2G to 3G systems and from 1G to 2G systems before that. There are many different views of what the fourth generation (4G) of mobile systems should look like. There are some characteristics that are common to many of these visions, including the following:

- *Order-of-magnitude bandwidth increases*. 4G systems probably will have data rates higher than those of 3G systems, by an order of magnitude or more. Since 3G systems in theory can provide data rates up to the order of 2 Mbps, and proposed incremental improvements go up to 10 Mbps, 4G bandwidth may need to provide tens of Mbps or hundreds of Mbps. However, there is a lot of debate about the need for such kind of bandwidth on wireless systems. Proponents of such data rates could point to similar increases in the memory storage required for PCs. Memory requirements always go up, and similarly, communications bandwidth requirements always go up. That new applications will need higher data rates is an evolutionary force that cannot be stopped. The new multimedia applications will demand high data rates. On the other hand, conservative skeptics claim that the costs of providing the additional bandwidth will not justify moving in that direction. Most of the time, when one is mobile (and that is when wireless is most important), one does not need full-motion, full-screen video.
- *New terminals*. Terminals will need to be multimode. One move in this direction is the current work on WLAN and 3G integration. Customers do not appreciate having to carry many communications devices, so an integrated device that connects through many access networks would be very attractive. Other ideas for 4G terminals include virtual reality terminals or other new user interfaces. Traditional user interfaces like keyboards and keypad may be less convenient than user interfaces relying on some combination of voice, movement (e.g., head turning or movement of other body parts for virtual reality terminals that might use a "goggle" type of terminal worn over the

eyes), or other data. Furthermore, recent developments in wireless sensor networking suggest that many future terminals may be sensors or embedded devices that do not interact directly with humans, and so may not require human user interfaces (this will be discussed further in Section 13.4).

• *Access-independent converged IP-centric network.* The 4G network will have an IP-based core that is accessed in multiple ways, as seen in Figure 13.1. Services will be access-independent, but will adapt to the particular characteristics of the access networks such as bandwidth and latency. The converged IP-centric network will provide global roaming and smooth handoffs between different access technologies. Thus, the WLAN and 3G integration work that is in progress can be seen as one step in this direction. Security issues would need to be carefully handled, but with the emergence of the AAA infrastructure and related technologies, IP is being evolved to handle the challenges.

• *New services and service-enabling platforms.* The 4G network will see a range of new multimedia services that will take advantage of the increased bandwidth available. Service negotiation would also be improved for 4G, and subscribers would be able to quickly purchase new services as desired, over the air. The rise of service-enabling platforms for the new network will help third-party application service providers to offer many new services. The initial stages of work in service-enabling platforms like Parlay and open service access (OSA) is already taking place, as discussed in Chapter 10. However, we expect that these platforms would only come into maturity for 4G. Furthermore, we may expect also to see increased levels of service integration, as mentioned in Chapter 1.

Meanwhile, mobility, QoS, and security support would be foundational components that are already being incorporated in 3G systems, but that may be enhanced in 4G systems.

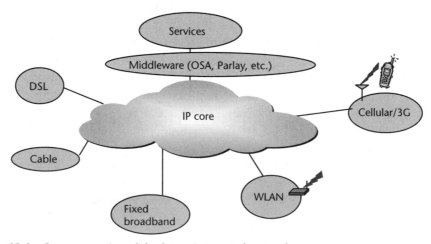

Figure 13.1   One conception of the future integrated network.

13.2 Technology Projections

Technology is rapidly progressing to enable some of the ideas mentioned above for 4G. A number of radio-transmission technologies have recently emerged that may be able to provide higher data rates and other features useful for 4G systems. These technologies are the subject of intense research, and they include orthogonal frequency division multiplexing (OFDM), multiple-input multiple-output (MIMO), and ultra-wideband (UWB). These will be introduced in Section 13.3.3.

Another area where technology is progressing is software radio (also known as software-defined radio). As multimode radios become increasingly necessary, it makes more and more sense to implement parts of the radio in software, because the cost of separate hardware radios for each kind of wireless technology becomes prohibitive. Advances in signal processing algorithm design, digital signal processors, and other enabling technologies are making software radio more attractive and feasible. In 4G, the concept of cognitive radio may also be explored. Cognitive radios are software radios that are "smart" in being able to avoid interference, as well as other in other functions. The term was coined by Joe Mittola, one of the pioneers of software radio.

In the network, IPv6 (Chapter 9) is slowly gaining momentum, and work continues on service-enabling platforms. Meanwhile, software technologies are evolving to handle more sophisticated applications in distributed environments, and where there may be power and memory constraints, such as in mobile devices.

13.3 A Complete Redesign or an Evolution?

3G systems like UMTS and cdma2000 are currently under development, with a few early deployments already in operation. Will 4G systems evolve from these 3G systems? We consider four schools of thought: (a) yes, 4G systems will evolve from these 3G systems, just as 3G systems evolved from 2G systems earlier, (b) no, 4G systems will evolve from 802.11 WLAN instead of from 3G systems, (c) no, 4G systems will be derived not from 3G systems or 802.11 WLAN, but from a different and newer wireless system, and (d) yes and no—4G systems will incorporate a variety of radio-access technologies, possibly including evolutions of the 3G radio interfaces, 802.11 WLAN, and other technologies.

We will discuss these four schools of thought in the sequence (b), (a), (c), and (d), in the following subsections. We discuss school (b) before school (a) because the view that 4G systems will be evolutions of today's 3G systems [school (a)], is in a sense the default path, barring any disruptions. After discussing school (b), where 802.11 WLAN is positioned as a powerful disruptive technology that could form the basis for 4G systems, we will be in a better position to examine the default path of school (a), having just seen a plausible alternative. The defense of school (a) will at that time be more interesting, thought-provoking, and fun!

13.3.1   The 4G Revolution Based on 802.11 WLAN

With its high data rates, inexpensive hardware, and deployment cost, and its use of cheap unlicensed spectrum, WLAN-based technology appears to have its cake and eat it too! Compared to 3G systems, WLAN provides order-of-magnitude-higher data rates, and costs much less to deploy. Furthermore, it is being proven in the marketplace as well. The rapid growth of the 802.11 WLAN market has been surprising to some, who expected it to fail or occupy niche markets at best, just like all previous attempts at WLANs had.

What made the difference? A good design and a workable system architecture certainly helped. However, I believe that three factors in 802.11's favor made all the difference in the world. The factors are standardization, placement in the IEEE 802 family, and good timing. Whereas previous WLAN systems were based on proprietary technologies and incompatible with one another, standardization of 802.11 allowed the devices to be produced by a number of different vendors, with fewer incompatibility problems. This allowed the number of devices to reach a critical mass and take on the catchy popular name wi-fi, further fueling public interest. Furthermore, 802.11 was not only standardized, it was standardized as part of the IEEE 802 family, which includes famous members like Ethernet (IEEE 802.3) and which shares a common logical link control (LLC) (802.2) over the MAC layer. This helped in the development of software drivers to work with 802.11 equipment, which could conveniently provide a similar interface to higher-layer software as 802.3 drivers. Given the large installed base of software designed to work over Ethernet, this was a real advantage. Another way of looking at it is that 802.11 did not come from nowhere and become an instant success—rather, it builds on existing technology (albeit with extensive modifications to support mobility and characteristics of MHs like the need to consume very little power). One could say that 802.11 evolved from Ethernet concepts, analogous to UMTS evolving from GSM. In fact, one of the informal names for 802.11 is wireless Ethernet. Finally, good timing was also a factor. The Internet had reached the stage where many people used it frequently and were eager to experience a mobile extension to their desktop experience.

13.3.2   The 4G Evolution Based on Improvements to 3G

It may appear that 3G systems are dinosaurs heavy with legacy-system support (backward compatibility) that adopt a network model not well suited for data traffic. The data rates provided are not very high compared with 802.11-based systems.

So, what do 3G systems have going for them? They are designed and built after years of experience in designing and building cellular systems from 2G systems; such experience can be carried over and adapted for the 3G systems. Even though there are significant differences between 2G and 3G systems, each of these systems is incredibly complex, with many layers of interactions between many different components. These systems handle mobility management, security features, and QoS, as well as network management, and provide many features, all in a well-integrated manner. On the other hand, 802.11-based systems, although growing rapidly in popularity, are still going through growing pains, plagued with problems handling security, QoS, and other features. Two main reasons for this are: (a) there is less experience with 802.11-based systems, since they are much newer and (b) coming as it does from the data world rather than the telephony world, 802.11 follows one characteristic of data networks in being traditionally less integrated than telephony networks (cellular systems came from the telephony world).

One may argue that these growing pains are only a phase, and that when we get over them, 802.11-based systems will be ready to serve as the foundation for 4G systems. However, cellular systems like 3G systems provide wide-area coverage, with cells that can be a few miles in diameter. The standard 802.11 WLAN is more appropriate for hotspot coverage than for blanket coverage of a city-sized area. By hotspot coverage, we mean coverage of areas (hotspots) where there is a high density of users, such as in airports, train stations, and hotels. The design of 802.11, like that of any other complex system involving many parameters and complex interactions between components, cannot be easily scaled to provide wide-area coverage. The use of unlicensed spectrum places severe constraints on the transmission capabilities of the system, for reasons including the need to be spread-spectrum based and transmission-power limitations. The CSMA/CA approach to medium access is based on local-area coverage, having similarities to the Ethernet MAC that is also designed for local-area coverage. For wider-area coverage, it becomes more efficient to handle medium access by dividing the wireless medium into channels and allocating the channels to nodes as appropriate.

13.3.3   The 4G Revolution Based on New Wireless Technologies

Some people would point out that both the cellular systems and the 802.11 WLAN systems have major shortcomings. The 4G systems would need to support high-data-rate data communications over wide areas. Cellular systems (including 3G cellular systems) do provide wide-area coverage but they support relatively low data rates by Internet standards given the explosion of broadband wired access in recent years. Furthermore, all the backward compatibility worked into the design of cellular systems has resulted in systems that have much baggage and are not well optimized for data. On the other hand, 802.11 WLAN can provide high data rates but is limited in range, partially because it is after all designed as a LAN technology, meant for a local area. The LAN-like MAC protocol bears some resemblance to the Ethernet MAC protocol.

One candidate for a 4G wireless technology is IEEE 802.16 wireless metropolitan area network (WMAN), popularly known as wi-max [1]. The standard 802.16 has some of the benefits of 802.11, such as being designed from the ground up to support data, unlike cellular systems that were designed for voice only at first and then later, in a sense, retrofitted to support data. However, 802.16 is not a LAN technology. The MAC is based on time slots, rather than a variation of a carrier-sensing MAC. Therefore, its operational range is further than that of 802.11 WLAN.

Furthermore, to provide higher data rates, MIMO technology and other smart-antenna technologies have emerged in recent years to exploit spatial diversity. Hence, just by adding antennas to both sides of the wireless link, the rate can be increased, *without* a corresponding increase in total transmit power or in bandwidth. This assumes a rich scattering environment, proper antenna placing, and other factors. However, the window is open for a leap in the bandwidth efficiency of radio links. We expect that MIMO would start to appear in the later versions of 3G systems (already, MIMO techniques are being considered in 3GPP standards meetings for Release 6), and be more mature and fully utilized in 4G systems.

Another candidate for a 4G wireless technology is orthogonal frequency division multiplexing (OFDM). OFDM is a way of transmitting signals on N parallel subcarriers, each at a slightly different carrier frequency (hence, the frequency division multiplexing in the name; orthogonality allows the signals on the different subcarriers to be separated for reception at the receiver). Central to the signal processing in OFDM is the use of the fast Fourier transform (FFT) and inverse FFT (IFFT) to demultiplex and multiplex the data in the receiver and transmitter, respectively. OFDM can reduce the harmful effects of multipath because the symbol rate on each subcarrier can be 1/N of the symbol rate on an equivalent single-carrier system. OFDM has proven to be a versatile technique in a variety of wireless communications systems. These include the widely deployed digital audio broadcast (DAB) and digital video broadcast (DVB) systems, and the 802.11a physical layer for wireless LANs.

A company called Flarion has applied OFDM to a new wireless system for high data rate voice and data communications [2]. They call the technology Flash OFDM, and they claim it is suitable to be the basis of the next wireless personal communications systems, since some of the technology problems with OFDM (like the peak-to-average ratio problem) have been recently solved, allowing the features of OFDM to shine. On the other hand, behind the Flash OFDM air interface, an all-IP network would be operating, optimized for data, but also handling voice by treating voice as "simply a persistent low-bit-rate form of interactive data" [3]. This system is being considered in the IEEE 802.20 standards group.

Yet another candidate for a 4G wireless technology is UWB technology. Ultra-wideband (UWB) technology can be thought of as an extreme case of spread-spectrum technology with many proposed applications in communications. Its characteristics include:

1. *Large bandwidths.* The transmission bandwidths employed by UWB systems is usually much larger than the transmission bandwidths of typical spread-spectrum systems, being on the order of gigahertz rather than megahertz. Even the wideband CDMA mobile systems being proposed as 3G cellular systems have transmission bandwidths in the order of 5, 10, or 20 MHz.

2. *Large fractional bandwidths.* UWB systems tend to have relatively larger *fractional bandwidths* than traditional communications systems. This concept is explained below.

UWB systems are commonly defined in terms of fractional bandwidth. Specifically, a particular radiation is considered ultra-wideband if it satisfies:

$$\frac{(f_u - f_l)}{(f_u + f_l)/2} > \frac{1}{4}$$

(13.1)

where f_u and f_l are the upper and lower frequencies (where the spectrum is 3 dB down or 20 dB down from the peak value, or between which 90% or 99% of the energy is contained, for example), and $f_u > f_l$. This rule of thumb is also known as the "25% rule" [4]. In contrast, even proposed wideband CDMA mobile systems will operate with fractional bandwidths on the order of 1%.

One key idea is that rather than try to avoid interfering with (or being interfered by) other radios, by using just a slice of spectrum (whether licensed or unlicensed), UWB uses a large chunk of the whole radio spectrum, on the order of GHz. What about interference? UWB exploits its large bandwidth to have very high interference rejection performance. Also, the UWB signal is spread out so thinly over its large bandwidth that the interference to other radios is negligible. Another key idea is that a carrier may not be required, meaning that UWB may be transmitted at baseband, allowing savings in power and complexity by eliminating many traditional radio components.

Since UWB usage fits into neither the traditional licensed nor unlicensed spectrum usage, regulatory agencies have been struggling to decide how to best regulate UWB usage. Currently, the FCC in the United States allows UWB communications provided they meet certain emissions criteria. Generally, these limits are the "Part 15" limits imposed on unintentional radiators, with additional notches in sensitive bands to protect systems like global positioning system (GPS). The original drivers of the regulatory process were companies pushing UWB using nanosecond pulses and pulse-position modulations. However, since these rulings have been made, alternative UWB schemes using OFDM and CDMA have also been proposed (see, for example, the Multi-Band OFDM Alliance's proposal) [5].

13.3.4   The 4G Evolution and Revolution Based on Heterogeneous Network Integration

Cellular networks provide wide-area coverage at high mobility and support moderate data rates. However, these rates are insufficient for many multimedia applications. WLANs support higher data rates, but cover smaller areas and support limited mobility. The complementary characteristics of 3G networks and WLANs have raised much interest in the integration of 3G and WLAN [6]. We can conceive of a multimode terminal that can access either WLANs or cellular networks, depending on what coverage is available where the terminal is located (and perhaps the user may be given the choice of which network to use, where applicable). However, this approach requires separate subscriptions to each network. Furthermore, even if the subscription and billing issues can be solved using AAA (which is indeed one level of integration of WLAN and cellular networks), seamless mobility between the two networks would not be supported unless additional integration features are added. In fact, 3GPP has defined six levels of integration between WLAN and cellular networks.

Integration of WLAN and cellular networks can be handled in different ways. A popular way of classifying these alternative network architectures is to divide them into three classes, namely, tight coupling, loose coupling, and no coupling. Since the definitions of these terms are still evolving, we provide our definition of the terms. When networks are coupled, we mean that they appear as one network, from the network layer and above of the protocol stack. One network is a master network and another a slave network, and the integrated network appears like a network of the type of the master network. Signaling and data from the network layer and above are what would be used for a typical deployment of the master network. The slave network provides link-layer transport between the mobile host and an interworking gateway network element that hides the fact that the slave network is used on the other side of the gateway. Typically, the gateway network element emulates one of the master network elements.

With tight coupling, all data and network-layer-and-above signaling traffic going to and from the slave network pass through the master network. With loose coupling, some network-layer-and-above signaling traffic going to and from the slave network may pass through the master network—however, the data traffic need not pass through the master network. The main motivation for tight coupling is to facilitate, as much as possible, reuse of the protocols of the master network from the network layer and above. One drawback of tight coupling, however, is that the master network may turn out to be a bottleneck for data traffic, for example, if the master network is a UMTS network and the slave network is an 802.11 network. Loose coupling attempts to address the problem of tight coupling, since data traffic can bypass the master network. However, it may require additional security gateway and firewall features in the network.

The no-coupling approach, on the other hand, treats the two networks as peers. There is no attempt at vertical integration of the two networks. However, interworking is assisted by the use of an AAA infrastructure.

The subscriber only needs a single subscription. For the master/slave architectures, the subscription is with the UMTS network and will enable the subscriber to use WLAN access without requiring an additional subscription. For the no-coupling (peer network) architecture, the subscription may be with either.

13.4   A Transformation of Wireless IP Devices

The field of wireless-sensor networking has been explosively developing in recent years. Wireless sensors found a place in the February 2003 edition of *Technology*

Review as one of the 10 "emerging technologies that will change the world" [7]. While much attention has been focused on networking of computers and mobile phones (via the Internet, as well as other protocols) over the last 20 years, recent forecasts indicate that future networks will be dominated by small and embedded devices. While the typical device on the Internet today is some form of PC (e.g., desktop or laptop) or mobile phone, the typical device on the Internet tomorrow would be a sensor or other small device, perhaps embedded in an appliance. Networked sensors are an increasingly important segment of the small- and embedded-devices arena. It is predicted that there will be 2.5 billion devices on the Internet by 2006, mostly mobile phones and terminals, and that there will be 60 trillion wireless sensors by 2010 [8, 9].

Researchers have begun exploring wireless-sensor networking in depth over the past couple of years. Research is active in many areas, including modulation schemes (e.g., is UWB good for sensor networks, or is something more traditional better?), design of power-efficient hardware, suitable MAC protocols that are energy efficient, routing protocols (e.g., power efficient, datacentric, and perhaps attribute-based addressing and location awareness), data aggregation, novel data dissemination algorithms, and application-layer query and dissemination protocols.

Thus, 4G networks may see an interesting paradigm shift from 3G networks—where each 3G end-user device (the same applying for 2G and 1G phones too) is usually associated with one human user, the typical 4G device may not have that kind of attachment. This has interesting implications. For example, GUI design will be less of an issue for devices that do not interface directly with humans. In a smart-home network with dozens or hundreds of networked devices, perhaps only a few might need to have GUIs for interaction with humans.

13.5   Summary and Conclusions

The future continues to look bright, and new wireless and Internet technologies continue to be developed at a fast pace. In this chapter, we look ahead, particularly at what might be the next generation of wireless systems. While there are many visions of what 4G entails, some elements common to many of these visions are order-of-magnitude bandwidth increases, new terminals, access-independent converged IP networks, and new services and service-enabling platforms. We examine some of the technologies that might play a part in 4G systems, including MIMO, UWB, and OFDM for the physical layer, software radio, and IPv6. We consider four scenarios for 4G: based on evolution from 3G systems, evolution from wireless LAN systems, integration of different technologies like 3G and wireless LANs, and a new technology.

Regardless of the shape that the next generation of wireless systems takes, there is generally a consensus that IP will play a big role in the network.

References

[1]   Eklund, C., et al., "IEEE Standard 802.16: A Technical Overview of the WirelessMAN Air Interface for Broadband Wireless Access," *IEEE Communications Magazine,* June 2002, pp. 98–107.

[2]   http://www.flarion.com.

[3]   Corson, M. S., et al., "A New Paradigm for IP-Based Cellular Networks," *IEEE IT Professional,* November/December 2001, pp. 20–29.

[4]   OSD/DARPA, Ultra-Wideband Radar Review Panel, R-6280, "Assessment of Ultra-Wideband (UWB) Technology," Arlington, VA, July 1990.

[5]   http://www.multibandofdm.org.

[6]   Varma, V., et al., "Integration of 3G and WLAN," *IEEE Communications Magazine*, Special Issue, November 2003.

[7]   "10 Emerging Technologies That Will Change the World," *Technology Review,* February 2003.

[8]   Cerf, V., "Internet in the Next 5 to 10 Years," available at http://global.mci.com/de//resources/cerfs_up/issues/internet_in5to10.xml.

[9]   Walrod, J., "Sensor Network Technology for Joint Undersea Warfare," *NDIA Joint Undersea Warfare Technology Conference*, San Diego, CA, March 2002.

About the Author

K. Daniel Wong received his B.S.E. (with highest honors) from Princeton University, Princeton, New Jersey, in 1992, and his M.Sc. and Ph.D. from Stanford University, Stanford, California, in 1994 and 1998, respectively, all in electrical engineering. He received the G. David Forney, Jr., Prize from Princeton University, the Telcordia Technologies CEO Award in 2002, and is also a member of Tau Beta Pi, Sigma Xi, and Phi Beta Kappa. He has been a research scientist at Telcordia Technologies in New Jersey since 1998, where he has been involved in research, development, design, and prototyping in wireless communications systems. His research interests include mobility management for IP networks, ad hoc networks, cellular networks, network security, physical layer technologies for wireless communications (e.g., UWB and OFDM), and WLAN/3G integration. Since August 2003, he has also been an assistant professor at the Malaysia University of Science and Technology (MUST). Dr. Wong is a board member of the Sister Societies Board of IEEE Communications Society, and a member of the editorial board of *IEEE Communications Surveys and Tutorials*. He is a Senior Member of the IEEE and has published and presented papers and taught tutorials in various IEEE conferences. He has also organized and chaired sessions at these IEEE conferences, including Globecom 2002, WMPC 2002, Milcom 2002, WCNC 2003, and WCNC 2004. He was a technical track cochair for the Multimedia Communications track for ICCCAS 2004. He contributed a chapter on IP mobility management in the book *Wireless IP and Building the Mobile Internet*, edited by Sudhir Dixit and Ramjee Prasad (Artech House, 2003). Dr. Wong has served as a guest editor for an *IEEE Communications Magazine* special issue on "Integration of 3G Wireless and Wireless LANs." He holds several patents and is listed in Marquis' *Who's Who in America* (2003–), *Who's Who in Science and Engineering* (2004–), and *Who's Who in the World* (2004–).

Index

Recent Titles in the Artech House
Mobile Communications Series

John Walker, Series Editor

Wireless Communications in Developing Countries: Cellular and Satellite Systems, Rachael E. Schwartz

Wireless Intelligent Networking, Gerry Christensen, Paul G. Florack, and Robert Duncan

Wireless Internet Telecommunications, K. Daniel Wong

Wireless LAN Standards and Applications, Asunción Santamaría and Francisco J. López-Hernández, editors

Wireless Technician's Handbook, Second Edition, Andrew Miceli

For further information on these and other Artech House titles, including previously considered out-of-print books now available through our In-Print-Forever® (IPF®) program, contact:

Artech House
685 Canton Street
Norwood, MA 02062
Phone: 781-769-9750
Fax: 781-769-6334
e-mail: artech@artechhouse.com

Artech House
46 Gillingham Street
London SW1V 1AH UK
Phone: +44 (0)20 7596-8750
Fax: +44 (0)20 7630-0166
e-mail: artech-uk@artechhouse.com

Find us on the World Wide Web at:
www.artechhouse.com